The Insect Epiphany

THE INSECT EPIPHANY

How Our Six-Legged Allies
Shape Human Culture

BARRETT KLEIN

Timber Press
PORTLAND, OREGON

Frontispiece: Art by Elizabeth Jean Younce

Timber Press
Workman Publishing
Hachette Book Group, Inc.
1290 Avenue of the Americas
New York, New York 10104
timberpress.com

Timber Press is an imprint of Workman Publishing, a division of Hachette Book Group, Inc.
The Timber Press name and logo are registered trademarks of Hachette Book Group, Inc.

Printed in China on responsibly sourced paper

Text design by Sarah Crumb
Cover illustration by Elizabeth Jean Younce; case design by Sarah Crumb
Illustrations on pp. 2, 23, 179, and 247 by Elizabeth Jean Younce

The publisher is not responsible for websites (or their content) that are not owned by
the publisher.

The Hachette Speakers Bureau provides a wide range of authors for speaking events. To find
out more, go to hachettespeakersbureau.com or email hachettespeakers@hbgusa.com.

The Central Intelligence Agency has not approved or endorsed the contents of this publication.

ISBN 978-1-64326-136-2

A catalog record for this book is available from the Library of Congress.

There are bastions of miniature marvels all about us.
Most are unknown to us, and their inherent value and cultural connections
will never be realized. I dedicate this book to the insects, and to the people
whose cultural traditions take time to appreciate them.

AUTHOR'S NOTE

The spellbinding diversity of insects is complemented by a diversity of humans and cultures who have developed unique approaches to working with, making, or becoming insects in boundlessly creative, stimulating ways. I strived to phrase facts accurately and express ideas respectfully, though some material is drawn from sources whose philosophy deviates from my own. If names or terms do not honor the people or their traditions as semantics evolve, my intent is to project a deep admiration for all who have formed personal relationships with underappreciated organisms. Many cultures are not mentioned in this book. This does not reflect a lack of insect connections, but an unintentional bias in my sample that I hope to reduce as I continue to explore the countless cultural bonds we have forged with our arthropod allies.

For additional materials and updates, please visit pupating.org.

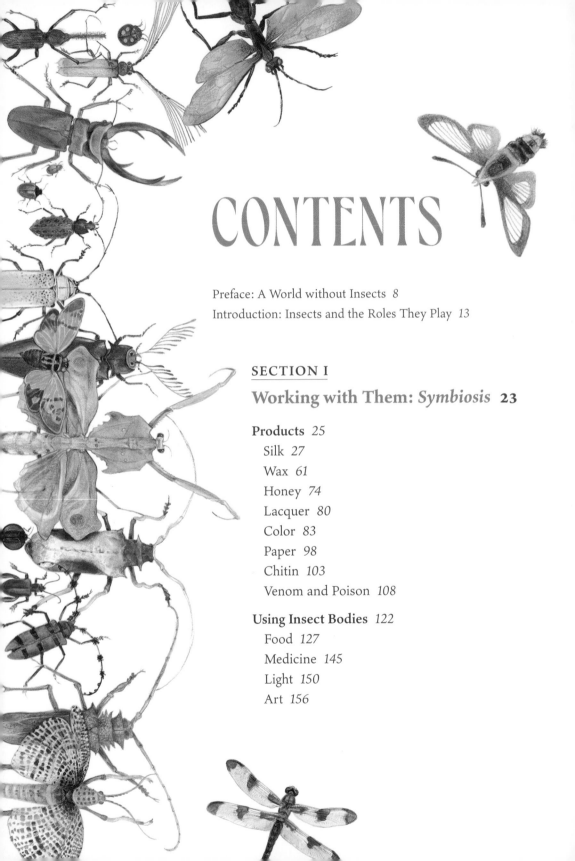

CONTENTS

SECTION II

Making Them: *Genesis* **179**

SECTION III

Becoming Them: *Metamorphosis* **247**

A WORLD WITHOUT
Insects

What does it take to spark a child's passion for the natural world?

A dead butterfly caught my eye, halted my play on a bright sunny day, and filled me with a curious exhilaration. I was five, living on the outskirts of Detroit, had no experience with wild areas, and only the simplest connection to the animals around me. But here, in my hand, was the delicate body of a once-living animal. I admired the intricate parts—the jointed legs, coiled tongue, bulbous eyes, beaded antennae, and vibrant colors. Through the tragedy of such a beautiful organism's death, I had received a gift, and took advantage of the opportunity by examining every detail. The body was not as fragile as I imagined it would be. I could stretch the legs and open the wings, and obtain a sense of how this individual had recently lived and moved. Here was a natural marvel that deserved my attention and my care. What else was out there? My mind whirled and my body tingled with electric anticipation that sustains me even today. I was a little boy in my backyard, and it was as if the world of natural wonders had suddenly become accessible to me. Though I had no idea what the future would look like, one thing was certain—my life had to be full of insects.

As I grew, a small world of life darted about in the dark water of a pond near my grandparents' cottage. I'd dip in a net, pull out a tangle of slimy green, and sift through the goo for surprises. I discovered my first water scorpion, predaceous diving beetle, giant water bug, and immature dragonflies with their unfolding labial masks, able to snare fish in a flash. I set up tanks and watched them hunt, grow, transform, and reproduce. A fishing spider held a sac of eggs under her body for weeks until I returned home from school and found spiderlings speckling the wall that led to my parents' bedroom. Until my mother reads this, she will have had no idea that I spent that afternoon carefully gathering each of the hundreds of wee silk-spinning offspring, and finished minutes before her return from work. Decades later, now a professor, I teach about insects and their relatives. I study their behavior, trying to understand what happens when they sleep.

◆

The spark ignited by that butterfly continues to grow as I learn how little I know. It was only much later, for example, that I learned how insects affect all of our lives, often in hidden ways. Quantifying this impact is not easy, but one way to recognize insects' significance is to imagine their absence. Try the following:

Close your eyes and picture your life exactly as it is, but in a world devoid of insects. How different would it be? Elements of this picture might put your mind at ease. No buzzing fly or jumping flea could spread disease or discomfort. No sucking bug or chewing caterpillar would diminish our crops. No wasp stings, no ant invasions, no cockroaches in the dead of night, and buildings would be spared the munching onslaught of termites. The whirring, creeping, and biting would all be vanquished—our days cleansed of the hordes, the legions of relentless invaders. Life would suddenly seem less harried, less troubled. We could go about our day less encumbered, with ten quintillion fewer things to worry about.

But let's examine that life more carefully.

It is true that some human diseases would disappear and sectors of agriculture would be less blemished, but only a tiny fraction of insects spread disease or compete with us for our crops. The absence of the remaining quintillions would be felt so profoundly that the dream of an insect-free life of peace would crumble in the face of collapsing ecosystems. We would realize that an interconnected web of organisms is a fragile entity, and if the most diverse animal lineage (insects!) were removed from this web, the effect

on others would be catastrophic. Songbirds reliant on insects would be silenced. Many fish, amphibians, reptiles, and mammals would quickly suffer and perish, to list only animals with backbones. We would likely join that unhappy lot, despite our aspirations for separateness from other animals. This insect-free world would no longer offer the fruits, seeds, and other insect-pollinated crops on which we depend. Busy ants disperse nutrients used by plants, and their digging brings air and water to the plants' roots. Insects and others decompose the world's bodily wastes.

Without insects, we would not know the taste of honey or the feel of silk. No butterflies to flutter by, no glistening jewel beetles, castles of clay built by termites, or nocturnal flashes of fireflies. Engineers would have fewer sources of inspiration to solve problems, and artists would lose their insect muses. Medicine and our knowledge of genetics would look very different without lessons learned from one species of fruit fly. Insects can indicate to us how clean our rivers are, or help us solve murder mysteries. Insects have affected the global economy and the outcomes of wars. They serve as religious symbols, fictional protagonists, metaphors, icons, and mascots. Our cultural heritage—what largely distinguishes and defines us—is inextricably linked to insects. Insects affect how we dress and speak, what we eat, where we live or travel, and how we perceive the world.

2 cm

The scope of cultural entomology, a field of study that examines ways in which insects influence human culture, is so vast that no book could do justice to its breadth or depth, and any approach would bias the story of our relationship with insects. My aim is to compel curious readers to explore more deeply and in a personal way their emotional connections or intricate alliances with insects. Though I exclude most non-insect lineages of life out of necessity, there is one group for which I make an exception. The arachnids differ in important ways from insects, but affect human cultures dramatically enough that I knew better than to ignore them. Spiders weave a culturally compelling tapestry throughout human history, and these are the primary arachnids we will see appear alongside their six-legged cousins.

The oldest known depiction by a human of an insect, an engraving of either a cave cricket (*Troglophilus* sp., family Rhaphidophoridae) or a katydid (family Tettigoniidae) sits among four (hints of) birds on this piece of bison bone found in the cave of Enlène, France. Look for the jumping hindlegs (*top* and *slightly left of center*).

To help accommodate a wide range of culturally important, bizarre, or thought-provoking topics, I divide this book into three broad sections. "Working with Them: *Symbiosis*," the first section, tells of ways in which we use insects' bodies, inside and out, for our cultural practices. We start by diving into insects' innards to reveal how silk, wax, and other bodily products important to an insect's survival have become useful to our own. "Making Them: *Genesis*," the second section, brings engineers, spies, architects, and scientists together in their utilitarian efforts to mechanically mimic an insect. "Becoming Them: *Metamorphosis*," the third section, highlights ways in which we imitate insects, or in some cases, ways in which it could be wise to do so. Each section offers evidence in support of one inescapable conclusion: Insects play a fundamental role in what makes us human. For some, this idea might bring discomfort, but for me it is an epiphany with promise. We can bridge our bodies, our histories, and our lives to the enthralling communities all around us. We can revel in knowing we are deeply connected to our multifarious and multifaceted neighbors. We can choose to celebrate insects, knowing that without them we would sacrifice significant aspects of our heritage, our humanity, and much of life as we know it.

Insects

AND THE ROLES THEY PLAY

Every living being begins as one cell. Many remain in this state, protected by a double layer of fatty acids that separate the outside world from what they maintain inside. Their entire identity is based on this barrier and what it holds. Others begin to divide, multiply, and develop into multicellular entities. Having more cells usually means growing larger, which can reduce risks of being killed by others. More cells also means more opportunities to partition a body to perform different tasks simultaneously. When multicellular, it can help to have an up and a down. Sometimes, it can help to also have a front and a back. For most animals, the front develops into a distinguishable head, which houses structures used to sense what is nutritious or dangerous when moving through the environment. Some develop a segmented body, with repeated body segments serving as functional backups. If one of the many segments of a segmented worm experiences neural failure, other segments may carry on. Segmentation can also bring about regional specialization, so that one body region handles food while another processes a previous meal. Pairs of appendages develop on segments within several lineages, and for more than 73 percent of all known animal species, appendages take the form of segmented, jointed limbs. These are the arthropods—arachnids (spiders, scorpions, ticks and mites, etc.) and fellow chelicerates, myriapods (centipedes and millipedes), and crustaceans. From within the crustaceans, insects arose to become the most diverse group of organisms documented on Earth. Six legs, external mouthparts, and a special sensory organ inside a pair of antennae begin the story of what it means to be an insect.

The earliest insects, four hundred million years ago, were wingless, and this lifestyle has served their descendants—jumping bristletails, silverfish, and firebrats—quite well. The introduction of wings, however, changed everything. Flight meant newfound abilities to find food, to escape predators, or to disperse more broadly. As entomologist and author Michael Engel put it, "Insects gave the world flight, and flight gave insects the world." Insects were the first animals to fly, taking advantage of life on the wing long before pterosaurs, birds, or bats. Nearly three hundred million years ago, the largest insects ever to have lived were aerial predators. Griffinflies, with wingspans greater than some of today's falcons (28 in. / 71 cm), shared many features with our dragonflies, damselflies, and mayflies. As "old-winged" insects, their wings stretched out, exposed and unable to fold over their backs. Other insect lineages overcame this limitation with the evolution of a tiny armored plate. This anatomical tweak spearheaded insect domination. By swiveling and collapsing, folding, packing, or concealing their wings, insects could burrow, swim, and slip into nooks and crevices, filling countless niches on land and in freshwater.

Wings do not come fully formed, and the manner in which they develop is another key to insects' success story. Wings can develop as small pads that become larger with each molt until adulthood, or they can dramatically unfurl after a succession of stages during which any hint of future flight is concealed. This is the hallmark of fully metamorphosing insects, in which the life cycle is no longer egg-nymphal stages-adult, but egg-larval stages-pupa-adult. The transformation from creeping caterpillar to winged butterfly is due to a strangely immobile and fixed pupal stage, during which almost every system of the body liquifies and reconfigures. The immature insect prepares for a completely different lifestyle, which means that immatures no longer need to compete with their parents for food. The transition is full of risks, but the value of going through the transformative experience is obvious when you compare the number of insect species with a pupal stage versus the number of species without. The silent, still pupa of an ant, fly, beetle, moth, or lacewing is never far away.

No one can tell you how many insect species there are, but among all species we have documented—from fungi to flying foxes—insects dominate. Robert May, a biologist who attempted to grapple with the diversity of animals, wrote "to a good approximation, all species are insects!" In less hyperbolic terms, about 60 percent of identified animal species are insects (1.1 million out of 1.8 million). The remaining 40 percent of species includes everything with a spine (less than 5 percent), and all of the non-insect multitudes without, including sponges, cnidarians, worms (flat, round, and segmented), mollusks, and others. Arachnids account for 6 percent of these "others."

If you slice a pie (chart) of life into the described species of each major lineage of organisms, insects (represented by a moth from Madagascar) dominate.

As for abundance, estimates are necessarily crude, but jaw-dropping. A locust swarm can contain one billion grasshoppers. A marching raid of driver ants can consist of fifty million workers, and the total number of ants on Earth may exceed twenty quadrillion. Pile just the carbon atoms making up all these ants on a scale, and that stack would weigh more (twelve million tons) than all of the world's wild birds and mammals put together, and equal 20 percent of the dry carbon mass of all humans. Throw in all the other insects, and they may number 1.25 billion for each human, and ten quintillion (10,000,000,000,000,000,000) total, give or take a magnitude or so. This vast workforce is so vital that to stand in the way of insects, to destroy their habitats, to squash or spray them out of hand, or to deny their importance, comes with consequences.

Insects have established themselves in every major habitat but the ocean. They are so fully integrated into Earth's ecosystems that their life habits can be viewed as services on which much of life depends. Entomologist and conservationist E. O. Wilson referred to insects as "the little things that run the world" because we depend on them to perform basic ecosystem services. Insects are nutrient, mineral, and waste recyclers, seed dispersers, tillers of the soil, food for others, and agents of population control, preventing unsustainable explosions of plants and animals. They are also pollinators.

If your food source resides in a flower, you will inevitably be covered in pollen grains, vehicles for the flowering plant's male sex cells. Continue to forage and you will unwittingly transport these grains to the female structures of other flowers. Like some bats and birds, insects serve as reproductive agents for flowering, fruiting plants. More than three-quarters of global food crops, accounting for one of every three bites we consume, depend on animals as pollinators, and insects are responsible for most of this pollination. So important are their pollination services that humans manage (by promoting or domesticating) at least twenty-two species of insect pollinators: two honey bees, nine bumblebees, eight solitary bees, and three flies (greenbottle fly and two hover flies). Many thousands of species of insects belonging to a handful of orders are wild pollinators. Think of an insect when you consume almonds, apples, apricots, blueberries, cardamom, cherries, coffee, coriander, cranberries, figs, grapes, mangoes, melons, nutmeg, papayas, peaches, pears, peppermint, pumpkins, raspberries, sesame seeds, strawberries, tomatoes, and vanilla. If you love chocolate, give thanks to the many unknown, minute insects responsible for pollinating cacao.

Insects increase our diversity, quality, and abundance of foods. We rely on their love of nectar for not only many of the fruits, vegetables, nuts, seeds, and oils we eat, but also some key ingredients for medicines, biofuels, and fibers. To picture what life lacking insect pollinators can be like, visit a remote community in the Hindu Kush Himalayan region of China. Subsistence farmers in this region have taken to stripping tracts of land and planting a more lucrative crop of apple trees. By removing the natural habitat of pollinating insects, spraying pesticides that kill pollinators, and planting only the most favorable fruiting trees and too few of the trees that are needed to cross pollinate these trees, the farmers have had to cope with low yields. Some farmers in the region have resorted to grafting branches of pollinator trees onto their favored trees, or attaching blossoms in bags of water to the trees, or bringing in hives of honey bees to replace wild bees, flies, and other pollinators. In the case of some farmers in Maoxian County, China, their solution has been to perform the jobs meant for insect pollinators themselves. What could have been a low-maintenance, natural process was transformed into a laborious procedure of employing seasonal workers to climb trees and dust pollen prepared from pollinator trees into each and every flower, top to bottom, tree by tree. Our destructive deviation from nature creates absurdities as unnatural as farmers armed with chicken-feather dusters role-playing as insects.

Attempting to estimate the economic value of domesticated or managed insect pollinators is fraught with unknowns and difficult to model, but even conservative estimates spell billions of dollars annually for the United States alone. Unmanaged, wild pollinators are also responsible for several billion dollars in the United States each year, and this is only accounting for production of fruits and vegetables. Back in 2006, entomologists John Losey and Mace Vaughan attempted to estimate a fraction of the economic value of insects by looking at a few of their ecological services, and their estimate

came to $60 billion annually for the United States. We know insect pollinators are important, but monetary figures tend to compel some people to act when they otherwise would not. For that reason, it can be worthwhile to calculate the incalculable value of insects to validate protecting them. When it isn't enough to speak up for the inherent beauty of a fly on a flower, with all of its history and nuanced symbiotic associations and environmental impact, we resort to defending its place in the world with a price tag.

Just as we label pollinating insects "beneficial," so do humans impose the label "pest" on other insects. Humans often employ the services of insect predators or parasites to avert crop losses caused by herbivorous insects competing for our food. The same is done to combat invasive insects, or vectors of disease. It can take one insect to reduce the abundance of another insect.

Insects are not restricted to the menus of other insects. Insects feed every vertebrate lineage, from fish to frogs and bats to birds, and they were

Flies, like this hover fly (family Syrphidae), are the uncelebrated pollinators of flowers, though in this case, the fly is feeding from an artificial flower that turns rain into sugar water and holds no pollen. Matilde Boelhouwer created *Insectology: Food for Buzz* (2018) to highlight the plight of urban pollinators, supplying the insects with food when natural resources wane.

undoubtedly consumed by early humans. Packed with protein, insects will play an increasing role as a human food source, potentially edging out the far less sustainable livestock on which we have long relied. Spiders and other animals without backbones also thrive on the throngs of insects, and insects are hosts for legions of fungi, and sources of nitrogen for carnivorous plants. Insects sustain the hungry masses, and without their diversity and abundance, ecosystems topple.

Another service on which we depend is so indecorous—so stinky—that unless it is our job to deal with it or the service is broken, we tend to avoid discussing or learning about it altogether. Every living being produces waste, and every living being dies, and we would be mired in feces and carcasses were it not for the valiant community of decomposers that operate largely underfoot. We are the only species that disposes of our bodily wastes by flushing them away to some distant place, typically for others to deal with. Though humans are remarkable in this feat of collective hygiene, the World Health Organization estimates that over 1.5 billion people lack basic sanitation, and 419 million of these people defecate in the open. For some traditional groups, what the World Health Organization considers "basic sanitation" may not be relevant or appropriate, but where people are densely packed, decomposing materials pose serious health risks.

We are not the only animals to benefit from the actions of decomposers. We bring many species into our lives, and it is easy to see how waste removal contributes to their well-being as well as ours. Livestock outnumber us, they vastly outweigh us (0.1 gigatons of carbon in all livestock versus 0.06 gigatons of carbon in all living humans), and since they share our propensity to poop, it is worth looking at the importance of removing dung excreted by some of livestock's biggest producers. A single cow produces about twenty thousand pounds (nine mt), or twenty-one cubic meters, of solid waste per year. With one billion cattle in the world, cows excrete the equivalent mass of forty-five million adult blue whales every year. That's forty-four million Statues of Liberty. With that much dung, we need decomposers to clear away the waste. For cows spending their lives on pasture or rangeland, and not raised on cement, dung beetles are efficient waste removers. Drawn to the scent of the bovine offerings, dung beetles fly from far and wide to extract their piece of the pie to feed their developing dung beetle young. The cow patty is a bustling, frenetic place of heated competition and thievery, so extracting a chunk of the ephemeral, valuable matter has to be done efficiently. Some burrow directly into the patty, others tunnel below and squirrel away dung from above, and still others form perfect spheres and roll the balls of dung from the heap. At least one species rolls their ball in a straight line away from the bustle by using polarized light of the sun or the moon as a guide. Egyptians were so struck by the rolling of a dung ball across the landscape that they ascribed divinity to the act. The dung ball was the sun moving across the heavens, and the sacred scarab beetle became Khepri, morning manifestation of the sun god Re. There can be no stronger expression of a people's devotion to decomposers than worshipping one as a god.

On a final fecal note, we can look at Australia's first cows to know what a landscape without insect decomposers looks (and smells) like. In 1787, the First Fleet sailed for 252 days from Portsmouth, England, bringing convicts, supplies, and livestock, including a small group of cattle, to Australia. After six months of being supervised, the tiny herd made a daring escape (or wandered off, as the case may be), established themselves in the wild, and multiplied. With introductions of more cows over the years, and a demand for dairy across the land, cattle quantities shot into the millions. As we know, millions of cows means voluminous and weighty wastes. These same cow patties would be dispersed and buried within twenty-four hours by dung beetles, were they deposited in South Africa. In Australia, however, cow patties decomposed slowly, festering in place, because few of Australia's more than five hundred native dung beetle species had evolved to work with dung resembling the wet excrement of cows. Australia is known for its koalas, kangaroos, wombats, numbats and bandicoots, and marsupials that produce coarser, pellet-like droppings. New decomposers were needed, and the Australian Dung Beetle Project (1965–1985), spearheaded by a Hungarian entomologist named George Bornemissza, introduced dung beetles adapted to processing cow dung, which quickly altered Australia's landscape. Removal

of dung by beetles meant more unfouled forage for cattle and far fewer parasites.

No human, no primate, no *mammal* has ever existed without insects on the planet. We rely on insects, whether we realize it or not, to perform tasks that keep our ecosystems going. Beyond offering essential ecosystem services, insects also affect the way we behave and believe. Cultural services performed by insects, though often difficult to quantify and frequently unrecognized, complement their ecological services in profound ways. Insects, as small as they are, feature largely in human culture and, starting with insects' bodily products, we can begin to appreciate how one tiny animal can influence billions. This is the tale of insect epiphany.

Rolling an unusually large ball of excreta through the London Zoo are two bronze *Dung Beetles* by Wendy Taylor CBE (1999).

WORKING WITH THEM: SYMBIOSIS

SYMBIOSIS

The night drops and shimmers
like a cloak of inky silk.
Why do you look to see life in the sky?
For travelers, abstract and enormous?
Hoping to see perfect symmetry,
fearful in its vastness,
galaxy to galaxy.
To see it would be to touch
the breathing back
of your sleeping partner
in the darkest
part of the night.
All the spinning emptiness
becomes a rise and fall,
numerical rhythm proving
you are not alone.
But bees, too, know these
beautiful mathematics.

They move from zero to creation.
The silkworm spins and spins,
like the planet you imagine contains
all the secrets of the universe—the answers
pulled out like an endless thread, translated
into something we can only feel.
Look at these gifts at your feet:
       honey, beeswax, silk.
From any other lover,
you would understand this devotion.
The bee hums and swoons around you,
an infinite galaxy, a dancing constellation.
Both of you bend to the center of the flower,
inhaling the illusion of time.

—ALISON ROGERS NAPOLEON,
   23 MAY 2023

PRODUCTS

Right now, your body is churning out so many molecules of ATP (adenosine triphosphate, the primary energy source to fuel your basic body functions) that even at rest, you produce half your weight in this tiny molecule every day. Your body, like a factory, is producing hormones, blood cells, and antibodies. You are replacing your own cells so often that your body is like a dynamic template rather than a static collection of tissues, organs, and organ systems. People have tried to place a monetary value on our material makeup, which ends up being a pittance if we are reduced to our chemical elements, but a fortune if we serve as organ donors. Humans have a strange, sometimes macabre, history of using our bodily fluids, wastes, or skin as ingredients in medicines, cosmetics, foods, or art. Binding books in human flesh—anthropodermic bibliopegy—was a practice not limited to the fictions of H. P. Lovecraft, Sam Raimi, or Peter Greenaway. "Mummy brown" is an artist's pigment that contained pulverized Egyptian mummies, until its corporeal source became better known, and mummy supplies dwindled. Medical history includes urine therapy, uses for breast milk, and the blood of dead gladiators to cure epilepsy.

Even today, researchers are making bricks out of human urine, and space food out of recycled astronauts' feces. All this may sound extreme, until we begin to think about what we eat and what we use on a daily basis that comes from other animals, or from fungi or plants. Survey the material culture in your life and you may find yourself surrounded by fibers, dyes, hides, adhesives, paper, and medicines derived from the bodies of our fellow organisms. Look at everything you consume and the variety of living and once-living sources that contribute to your diet. Our existence is tied to a diversity of

organisms, and their identities can be perfectly obvious, or hidden within indecipherable lists of ingredients or encapsulated within innocuous packaging.

Like all other living organisms, insects also resemble factories, and what they produce has the power to kill us, clothe us, or cure us. I will try to make clear the debt we owe to our six-legged neighbors by highlighting a few of the products most powerfully embedded in our cultures. Our reliance on insects as factories to synthesize useful or culturally enriching products means we must be mindful of insects' health and survival if we are to continue to depend on them. The loss of each species spells the permanent disappearance of everything bound within the body of that species, including possible answers to questions relevant to evolution and ecology, as well as to medicine and industry. What follows are products of insect glands that we use as food, clothing, adhesive, preservative, medicine, dye, or art medium. The insects making these products have reshaped international commerce, affecting our politics, our language, and our belief systems. The legacy of empires stands on the bodies of scale bugs, ancient art was cast from the offerings of stingless bees, and raising caterpillars erected a pantheon of goddesses. The world's distribution of wealth rests, in part, on the insides of insects.

Silk

The poet makes silk dresses out of worms.
—WALLACE STEVENS, FROM "ADAGIA"

The Silkworm

Dislodged from a high tree branch by a curious crow, a cocoon falls. Inside, a single plump larva, after weeks of eating, growing, molting, then building and transforming, is on the verge of molting again, this time to reveal legs, compound eyes, and wings. Plummeting from her high perch, the larva's quiet transition into a moth is interrupted by a soft landing in a hot cup of tea. Nothing in her species' evolutionary history prepared her for the heat that melts the glue binding the fibers of the walls that encase her. The fibers unravel, the influx of scalding liquid engulfs her, and her metamorphosis is cut short. The tea—her tomb—belongs to an empress, sitting in the shade of a mulberry tree in her imperial garden. Initially startled, Leizu is transfixed by the shimmering thread snaking on the surface of her beverage and takes note of the hapless victim, half-hidden within the dissolving cocoon. The thread continues to lengthen, and continues to glimmer. An idea crosses Leizu's mind as she has a lady-in-waiting pinch one end of the fiber and walk away. Down the garden's path, across the courtyard, beyond the fortress's walls, the servant finds herself at long last in the Forbidden City, far from the mulberry tree with the tiny teacup, the caterpillar, and the crow.

According to legend, Leizu, Lady of Xiling and primary wife of the Yellow Emperor Huangdi, was the founder of sericulture, the industry of rearing silkworms to produce silk. Leizu is thought to have been instrumental in the cultivation of mulberry trees to feed silkworm larvae, as well as in the rearing of silkworms, and reeling of silk from their cocoons. Legend further attributes to Leizu the invention of the loom, which could effectively weave the wares of worms into mass-marketed textiles, though the innovation may have actually come from an adviser to the emperor. True or not, the legend of Leizu offers a lesson. Paying attention to detail, to natural processes, can bring understanding, and potentially change the world. Here, it is about recognizing a natural substance and its source (silk and moth) for their value, intrinsic or otherwise.

The unwitting protagonist of this legend is *Bombyx mori*—the domestic silkworm moth—producer of every silk garment or commercial silk product you have likely ever touched or seen. This single species has dominated our attention and our silk-producing resources for roughly five thousand years. A silkworm uses glands her ancestors once used for producing saliva, and in a delicate and intricate feat, spins filaments from her head, which are glued together at room temperature into a single, unbroken thread, continuously wrapping herself in a cocoon. Unwound, that thread can form a fiber one mile long (300–900 meters), and stronger than most other natural fibers. As other fabrics degrade into powder, silk survives intact, evident in silk fragments dating from circa 2750 BCE. Silk is strong, but it is also light, smooth, absorbent, easily dyed, and a good insulator. Silk's primary allure is its sheen, and this it owes to a prismatic scattering of light caused by the fiber's molecular structure. Cut short, the threads are dulled, so traditional silk production uses a scaled-up version of Leizu's teacup to heat-treat the cocoons, preventing a new adult moth from snapping her special strand by killing her before she is allowed to emerge. The boiling water also dissolves the glue that holds the fibers together, releasing them to be reeled.

The source of this prized product, and the process of co-opting it for human use, was treated as a state secret, protected upon penalty of death in China. The profitable venture of exploiting moths was exclusively in the hands of the Chinese government until its secrets were eventually smuggled to silk-hungry neighbors. Smuggling

Duplicitous monks sneak silk out of China, spearheading the Byzantine Empire's silk trade (*Emperor Justinian Receiving the First Imported Silkworm Eggs from Nestorian Monks*), as depicted by Karel van Mallery (ca. 1595).

2.   Monachi duo Iuſtiniano Principi   Semen dedére, vermis vnde ſericus.

6.   Hinc vermium permulta ſæpe millia   Simul legunt, parantq́; telas feminæ.

Japan was an isolated nation when forced to open its doors to the world in 1853. Silk, which somehow reached Japan around the fourth century AD, transformed Japan into an international powerhouse within fifty years.

something as small as insects might sound easy, but when stakes are sufficiently high, state protections can be dauntingly prohibitive. This only means that smugglers had to be especially clever, as illustrated in the following two historical (apocryphal?) tales, presented here as ill-advised tutorials.

How to smuggle silkworms, strategy #1: Don an outrageously elaborate mélange of garments, too voluminous and precious for guards to search. Slip silkworm moth eggs into your ornate headdress and, with great pomp and confidence, leave the kingdom to establish your moth enterprise elsewhere. Note: This feat is more easily accomplished if you are a Chinese princess being married off to the King of Khotan.

Smuggling silkworms, strategy #2: Adopt the guise of extreme humility. Who would expect a pair of devout monks to engage in subterfuge? Stash silkworm eggs or young larvae in your bamboo walking cane and, after gaining the confidence of your hosts, depart with this treasure to the pleasure of Emperor Justinian I. Initiate a 650-year monopoly of Byzantine silk production in Europe.

Silk Goddesses and the Dog-Headed Woman

Silk brought prosperity, then it birthed gods. As can happen with traditional human practices and their founders, sericulture became a spiritual enterprise in China, and Leizu shed her mortality in the process. Followers worshipped and made sacrifices to a pantheon of patron saints in special sericulture shrines featuring statues of the sericulture goddesses. The deified Leizu stood among them as Silkworm Mother. Despite this high honor, it is not Leizu who draws our attention in certain depictions of the pantheon, but a goddess endowed with a second head—that of a horse. Springing from Chinese antiquity, the saga of the horse-head woman offers a strange silkworm metamorphosis origin story that could offer further insight into our bond with silkworm moths:

It had been a year since bandits had taken a man from his wife, from his daughter, and from the life he knew. Left in grief, the man's wife promised the marriage of their daughter, Can Nü, to anyone who retrieved her kidnapped father. Fearing the treacherous bandits, no man was up to the task. Only the man's horse—rearing, jolting, and breaking free of his halter—responded to the woman's plea. The horse bounded off, returning days later with the long-lost patriarch of the family. All seemed joyous, save for the father's confusion about his horse's strange new behavior. The horse, expecting his due matrimony, appeared driven and possessed, particularly whenever Can Nü drew near. An explanation was in order, but his wife's words did anything but allay the man's concerns. Not especially open to the concept of a marriage between daughter and horse, the father killed his horse with an arrow, flayed it, and left the hide to dry. Can Nü, walking past the remains of her father's rescuer, was enveloped by the horse hide and vanished, only to reappear in

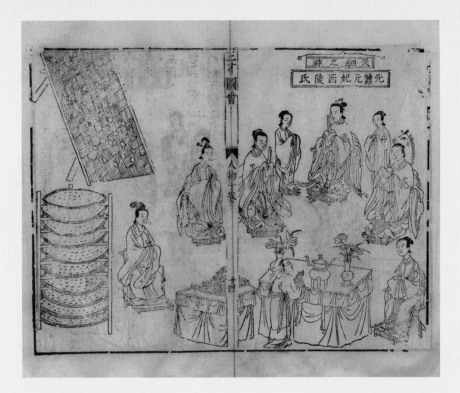

the guise of a silkworm. Her parents, distraught by Can Nü's transformation, watched as she spun a silk garment—her cocoon. Never before had they stopped to acknowledge the wonder of this commonplace behavior, but their daughter was now the larva, preparing to conceal herself in silk, and to transform. A newly metamorphosed Can Nü reappeared in the sky long enough to console her parents with the knowledge that she now and forever lives in heaven, rewarded with the dignified position of Lady of the Nine Palaces.

Leizu (嫘祖), and attendants, including the horse-headed silk goddess Can Nü (蠶女)

This tale leaves me with some unresolved questions. How did the daughter feel about the horse, or about her transformation? What power caused her to vanish and metamorphose? Is her newfound second head a neighing, functional addition to her body? As a parent, would I be consoled to learn that my daughter has become someone's additional wife for eternity? If nothing else, witnessing one's child turn into a caterpillar would make any parent look at insects in a different light. A lowly, leaf-consuming "worm" constructs her own shroud, within which a complete transformation ensues. Caterpillar to moth, daughter to heavenly entity, and China to silk-trading powerhouse.

So important was the production of silk to some parts of China that, for a lunar month, inhabitants were to avoid "silkworm taboos," consisting of social visits, celebrations, and other activities, so as not to upset the life cycle of the silkworms. China and the people of China became known to many as "Seres," a Greek term referring to silk. Their very identity was nominally tied to the product of a caterpillar. This precious offering helped fuel the Silk Road, a network of trade routes that connected the East and West for centuries. Trade extended across a continent, and China expanded its Great Wall to protect the lucrative commerce. Along with the shimmering fiber of a caterpillar came trade in other material goods. The Silk Road saw the spread of exotic foods, like cinnamon, coconuts, rhubarb, and turmeric, as well as the transport of horses and camels, or tigers and elephants. Peddled and preached, less tangible commodities also proliferated in the form of ideas, philosophies, cultural practices, and genes.

As with any road connecting people and their wares, the Silk Road also facilitated the spread of disease. The Black Death—a bacterial pandemic spread initially by rat fleas biting humans—raged through Europe, Asia, and North Africa. The Black Death caused the greatest wave of human death the world has ever seen. The rate of transmission was too horrifyingly fast for the rat flea to be entirely responsible, so explanations for the plague's devastation rely on human fleas subsequently carrying pneumonic plague, which could be spread by a human cough rather than a flea bite. There is some irony that trade invigorated by one insect indirectly facilitated a pandemic initially vectored by another insect. As one insect spread culture and commerce, another brought buboes and agony.

The effects of sericulture on human history have been tremendous, but however our lives have been shaped by the Silk Road, none of this can compare to the effects felt by the silkworm moths themselves. Let's begin by considering the ethics of sericulture from the moth's perspective. If one pound (0.45 kg) of silk requires roughly three thousand silkworm cocoons, and approximately seventy-five million pounds (34,019 mt) of silk are produced each year, enough four-inch (ten-cm)-long caterpillars are killed annually to circle the Earth's equator 561 times. Placed end to end, they would travel to

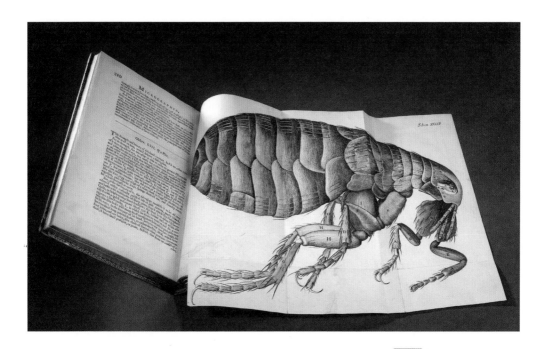

the moon and back 28 times. Four times as many caterpillars perish for our silk as all other animals combined
(including "shellfish") that die for human food in the
United States. The legacy of one cocoon in a teacup is
the annual, orchestrated massacre of billions.

Solution? Maybe we should resort to the age-old
inequity of reserving silk for royalty! Could we pay
penance, or acknowledge the silkworm sacrifice? The
Japanese perform an annual memorial service to insects
in the Silkworm Shrine (Kaiko no Yashiro) in Uzumasa,
Japan. Alternatively, what if we were to stop silk production altogether and
rally for silkworms' freedom? Liberation, as it turns out, would spell disaster
for the domesticated silkworm. As a demonstration of the moth's desperate
state of affairs, Tera Galanti creates art installations, like *Hope and Futility*
(2006), that place adult male moths on a platform situated slightly below a
platform of adult females. All the males need to do is to fly up to the females'
platform and they will achieve fitness victory. The males, tragically, never reach
the females. We have domesticated this species into a zone of utter reliance.

A foldout of a flea is one of Robert Hooke's explosive introductions to microscopy in *Micrographia: Or Some Physiological Descriptions of Minute Bodies Made by Magnifying Glasses with Observations and Inquiries Thereupon* (1665).

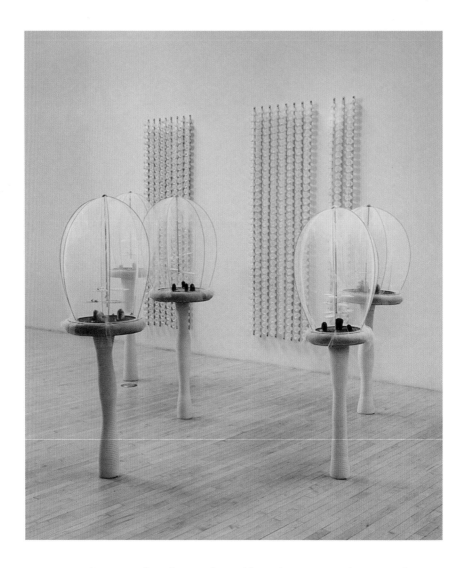

No *Bombyx mori* flies free in the wild. In fact, no *Bombyx mori* flies. Its closest relative is a larger-winged, wild silk moth, *Bombyx mandarina*. When compared with its flight-worthy relative, the domesticated silk moth is also deficient when it comes to perceiving odors, other than the chemical she uses to attract her mates. The domesticated moth is capable of mating with and producing viable offspring with its wild relative, but never the twain shall meet unless such a rendezvous is orchestrated by a human matchmaker. Galanti's art exposes the power we can wield when permanently modifying another species.

We have bred dogs and pigeons and cows and corn to appear and behave more to our liking, and we have artificially selected silkworm moths to suit us as well. We are familiar with the process of selecting for favorable traits and selecting against unfavorable traits in the foods we eat or the animals we breed. Charles Darwin used our familiarity with artificial selection to introduce his groundbreaking idea that natural selection could cause evolution when he wrote *On the Origin of Species* in 1859. A heritable trait, when favored (by something other than humans), can result in an individual's reproductive success. If competitors lack that trait, they will produce fewer offspring and become the losers in a struggle for survival. In the case of the silkworm moth, we have been breeding them, selecting for heritable traits favorable to us. Flight is normally an essential trait for the moths, but may have been viewed as dispensable or a hindrance to early sericulturists. Another possibility is that a mutation resulted in flightless moths and human handlers let the flightless trait continue to drift through the population of moths until the capacity for flight—for freedom—was lost.

As a consequence, we cannot free the silkworm moths and expect a happy transition to the wild for the species. They are bound to us if they are to survive. It is possible to collect silk without killing the developing moth by cutting the top of a cocoon and allowing the adult moth to emerge before reeling the silk, but the production devoted to the practice of using "nonviolent" silk is an exception to the sericultural norm. The practice still entails culling great numbers of moths at some stage of their life, and still results in flightless moths wholly reliant on humans for their survival. This inextricable bond between moth and human, although typically sustained in slaughter, also means that *B. mori* is far more successful in terms of its fitness. While other moths are declining in numbers, this domesticated species, which now owes its survival to humans, prospers under our management in terms of its abundance and geographic range.

This form of prospering under human control may seem like little consolation for the silkworm. The future of the silk industry could easily collapse, depending on our future fickle desires, or our survival. Feeding legions of insatiable caterpillars is expensive and involved, and produces a silk that is weaker than some others found in nature. An engineer's solution is to invent a superior synthetic surrogate to natural silk. Nylon and rayon are early examples of manufactured fabrics designed, in part, to replace silkworm silk. These and other silky substitutes differ in their attributes, but all are far from attaining the strength of natural silk. They also lack a surprising silk asset: the ability to vanish, without ash residue, when burned. This quality made silk a highly desirable material during World War I, when silk bags were used to hold powder charges for every big artillery shot ever fired by the United States. Some speculate that the very act that sparked World War I—the assassination of Archduke Franz Ferdinand of Austria during his visit to Serbia—could have been prevented had he worn a silk vest. Silkworm silk was a key ingredient of bullet-stopping body armor at the time, and the archduke may have owned one, but he took few precautions during his risk-filled and error-ridden fatal ride.

The fates of nations were tied to the domesticated silkworm. Japan was compelled to modernize or be colonized in the late nineteenth century, and raw silk became its primary export, transforming an agrarian society to a military power that led to victory in the Sino-Japanese War and entry into World War I. The silkworm's wartime role continued, as their silk played a part in the escape of Japan's enemies. Captured and incarcerated during World War II, British prisoners received unexpected gift packages containing board games. The games had hidden compartments covertly stocked with tools to facilitate a prisoner's escape. British intelligence officer Christopher Clayton Hutton included thin, silent, and durable silk maps detailing prisoners' locations relative to safe havens where they might find refuge. Somewhat ironically, the British used an alternative material obtained from a shipment en route to an enemy nation, Japan, to produce additional maps. The alternative? Paper pulp from silkworms' favorite food—mulberry leaves.

Exploding bags, bulletproof vests, and durable maps demonstrate silk's utility, but taffeta, tussah, chiffon, dupioni, georgette, habotai, organza, silk velvet, shantung, and crepe de chine conjure thoughts of luxury, and less of utility. For a time, the fabric in its different forms was reserved for royalty and nobility. It is now often associated with fine, traditional clothing throughout Asia, including the cheongsam, kimono, and sari, as well as with modern fashion pieces.

Sericulture (The Process of Making Silk), depicted on silk! Silk workers place cocoons in baskets in this small detail from a handscroll attributed to Liang Kai 梁楷 (early 1200s).

Being the versatile and resilient material it is, with a weight of history behind it, silk also looms large in the world of art, though just how extensively silk contributes to the history of art would be difficult to quantify. A search for "silk" in The Metropolitan Museum of Art's online database offers nearly forty thousand results. Entire museums, including the China

Words attributed to the philosopher Confucius are shrouded in the silk strands of silkworms in Xu Bing's *Silkworm Book: The Analects of Confucius* (2019)

National Silk Museum, are dedicated to silk. Silk might surpass any other insect product used in art, though no one has ever attempted to quantify such a comparison. One place to start such a daunting exercise would be to estimate the number of paintings on silk canvases in all the world's museums and galleries. Many, as it turns out, depict insects, and some even show the process of harvesting silk from silkworms.

In rare cases of contemporary art, silk and the silkworm moth take center stage. Since 1986, Kazuo Kadonaga has highlighted traditional practices of culturing silkworms. Viewers of Kadonaga's work can observe the process of tens of thousands of silkworms spinning, or the aftermath of their massive construction efforts. In one series, Kadonaga produced what appears to be a cityscape of crates exposing 110,000 silkworm cocoons, woven (and heat-treated) in place. Xu Bing has also put a spotlight on the caterpillars. Since 1994, Bing's conceptual art includes a series of works with live silkworms—on mulberry branches, on everyday objects like newspapers and a laptop, or on blank books in

110,000 silkworms roamed 91 pine and cedar grids, distributing themselves in response to the grids being turned for 48 hours by artist Kazuo Kadonaga and his assistant. Once settled, the caterpillars spun their cocoons and Kadonaga created a cityscape of the cocoon-filled structures in his studio.

which eggs resembling a cryptic script hatched and caterpillars emerged. In another, viewers could watch as silkworm larvae spun cocoons inside a video cassette recorder, set oddly beside a television playing a prerecorded view of the same.

Our last artistic silkworm collaborator for the moment is Jen Bervin, who created *Seven Silks* (2018), a study of silk in seven forms. Bervin harnessed the power of poetry, so alongside a vial of raw silk and a cocoon, she includes a medical innovation in the form of silk printed with microscopic poetry. This clear bit of silk, compatible with the human body, could be used to monitor a person's blood chemistry. As cutting-edge as this final inclusion may sound, any art piece that makes silkworm moths its subject is, ultimately, a piece about Leizu and her teacup, and the ancient practice of stealing a strand from its spinner.

Jar and vials harboring four of Jen Bervin's *Seven Silks* (2018), as well as a poem based on the silk genome

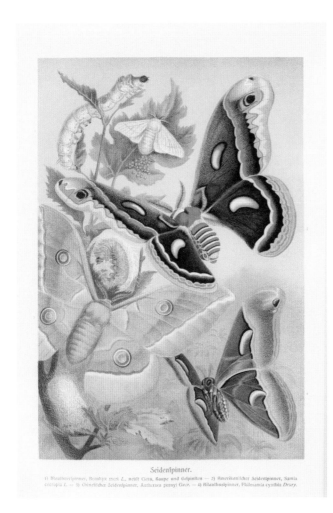

Silkworm moth
Bombyx mori
(pictured as cater-
pillar and adult on
the plant, *upper
left*) is joined here
by three other
silk moth species
(*Seidenspinner* 1915).

Seidenspinner.

1) Maulbeerspinner, Bombyx mori L., nebst Eiern, Raupe und Gespinsten — 2) Amerikanischer Seidenspinner, Samia cecropia L. — 3) Chinesischer Seidenspinner, Antheraea pernyi Guer. — 4) Ailanthusspinner, Philosamia cynthia Drury.

The 200,000 *Other* Species of Silk-Spinning Insects

The domesticated silkworm moth looms large in human culture, but there are more than 185,000 other species of moths and butterflies (Lepidoptera), and these spin silk as well. Though silk cocoons from some of these other species serve as ornament, rattle, or other use in various cultural practices, surprisingly few species have inspired entrepreneurs, engineers, or artists to

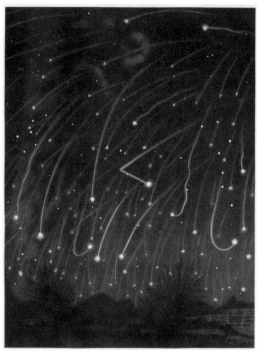

ABOVE LEFT: Traveling to a paper conservation conference in late spring of 1981, my mother drove from lush green forest to a brown and diminished Andover, Massachusetts, where none of the locals flinched as spongy moth caterpillars dropped like rain onto their outdoor banquet.

ABOVE RIGHT: With a future of woodland devastation looming, Étienne Léopold Trouvelot wisely redirected his attention from his failed introduction of an alternative silk-spinning moth to creating beautiful astronomical art.

exploit their wares. Some sericulture exists using wild species of moths, providing income and contributing to material culture in India, Madagascar, and elsewhere.

Seeking a viable alternative to introduce to the Americas, French scientist Étienne Léopold Trouvelot had the bright idea of working with a species of moth known to be outrageously opportunistic with respect to their dietary intake, eating more than five hundred types of trees and shrubs. Trouvelot transported egg masses from Europe to the United States in the 1860s with the goal of starting a sericulture industry. His choice, the spongy moth, *Lymantria dispar*, prospered in the United States, but not in the way he imagined. Larvae escaped from his backyard into the woods and spread far and wide, devouring forests in their path. The herbivorous rampage resulted in the loss of millions of hardwood trees and billions of dollars spent to control this invasive species. Trouvelot wisely changed fields, and dedicated the rest of his professional life to astronomy and producing a massive collection of astronomical art, as descendants of his caterpillars laid waste to eastern US forests.

Secreting silk is a common practice across many insect orders aside from Lepidoptera, and likely evolved multiple times within insects. Silk takes so many different forms that people avoid getting too specific about its makeup when defining it; if you can spin a filament from a liquid protein solution, you just made silk. To an insect, silk typically spells protection, particularly during times of defenseless transition. Webspinners (Embioptera) shoot silk from an enlarged segment in their front feet and form tunnels through which they dart forward and backward with ease, avoiding predators, foraging for food, and raising their young. Weaver ants (*Oecophylla* spp.) grasp legless larvae from their own colony and compel the immatures to spin silk, affixing leaves together to form arboreal nests for the entire colony. Even honey bee and bumblebee larvae spin silk to strengthen the cells in which they develop. You can find insect silk spanning the subterranean to the tops of trees.

Spinning silk also happens underwater. Caddisfly adults (Trichoptera) resemble moths, but generally have hairy, rather than scaly, wings. Their aquatic larvae spin silk to form nets, domes, hideaways, or cases that wrap around their relatively tender bodies. To form their cases, caddisfly larvae grasp and inspect particles that flow by or that they salvage from the floor of their aquatic environment, then bind the materials together. The silk is spun from a pair of modified salivary glands in their head. Instead of digesting food with what emanates from these glands, they spin a strong, sticky strand that can hold together rocks, sticks, mollusk shells, or other bits of matter into cases that can be species-specific. These quirky case-bearers may not seem noteworthy, but their very presence causes ripples across ecosystems. Caddisflies can be an important food source for others—vertebrate

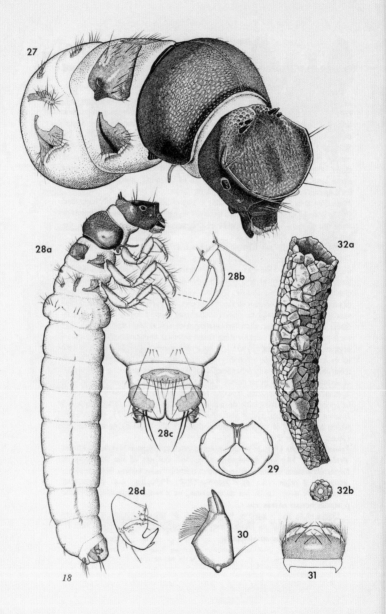

27

28a

28b

28c

29

28d

30

31

32a

32b

18

and invertebrate alike. From a human perspective, their aquatic larvae can help inform us whether bodies of water are relatively healthy or polluted. Forensic biologists have, in at least one case, used caddisflies associated with a drowning victim's remains to solve a case of homicide. Caddisfly larvae are used as fish bait, and a fly-fishing arsenal can include handcrafted, case-bearing caddisfly lures.

Of all the aquatic insects, caddisfly—and a couple of moth—larvae have been recognized as architects, but our concept of architecture is often rooted in design and less in construction. Do case-making caddisflies have a sense of design, or are they fixed in their innate species-specific ways of producing their works? As stereotyped as their building behavior is, caddisflies have to make do with the materials they have available when their favored material is inaccessible. No one has tested this more exquisitely than artist Hubert Duprat. Duprat relies on a caddisfly's flexibility when he prods a larva from a case, places the denuded larva on a bed of riches, and watches as the persnickety architect inspects gold spangles, pearls, and turquoise, and decides what passes the mysterious bar for case-making fodder. The larva rapidly draws the glistening substance through a pair of front legs and mouthparts and either summarily rejects it by tossing it aside, or inexplicably accepts it as worthy of pasting in place with silk. Duprat has long been fascinated by caddisflies, obsessively collecting historical interpretations, natural history observations, and cultural connections. Coercing them to produce mini bejeweled masterpieces seems a fitting means of intimately connecting with the insects.

ABOVE: It starts with a fashionable dress and a bagworm moth larva in need of case-making materials. Aki Inomata snips pieces, offers them to the larva, and a new garment is created using silk as adhesive.

Five new garments hang from a branch as the developing psychid moths pupate within.

Artist Aki Inomata has recently achieved a feat similar to Duprat's by offering bagworm moth larvae (family Psychidae) fabric snippets of haute fashion by well-known fashion designers. The larvae normally spin a protective case made of plant material, lichens, or soil to live in, but here they use their silk to dress themselves in material designed for humans. The act of offering fabric to would-be larval fashionistas follows a tradition among Japanese children of offering colored paper to caterpillars.

Whether you are a child or an adult, the trick of knowing how and when to stage a silk-spinning intervention depends on your knowledge of insect behavior. Hubert Duprat has become intimately familiar with caddisflies, and Aki Inomata must be familiar with the life cycles of bagworms. This helps the artists achieve something close to their intended goals. Though their works look like the inevitable outcomes of perfect planning, Duprat, Inomata, Kadonaga, and Bing must ultimately relinquish control of their art to the insects. Any artist wishing to work with live insects must be willing to accept unpredictable outcomes. Their six-legged associates will, usually, have the final say.

Spiders

Imagine secreting a material from your body that could be either spun, woven, used as a chemical-laden lure, snare, trapdoor, cocoon, net, nest, funnel retreat, mate attractant, or consumed. The same strand that could be used as an anchor could serve as a vehicle to whisk you away, take you aloft, and ride air currents to new sites. Depending on who secretes it, silk can spell safety, security, or predatory accessory. As mentioned in the introduction, spiders are not insects, but a discussion of silk would be woefully incomplete without the silks of spiders.

All 53,000-plus described species of spiders spin silk, using silk's various properties of strength, tension, elasticity, or stickiness as needed. Darwin's bark spider (*Caerostris darwini*) produces the toughest known silk—ten

A spider of dubious anatomy (legs attach to the front portion of a normal spider's body, though this could be a monstrous shape-shifting tsuchigumo of Japanese myth and legend) is perched on the back of a garment worn under a kimono made not of spider silk, but crepe silk from silkworms.

times tougher than Kevlar—toughness being the amount of work it would take for it to rupture. Yet, spider silk is so much less dense than other materials that it is stronger than steel on a per-weight basis. It's easy to see why poets, artists, philosophers, and engineers have been transfixed by the intricate and varied webworks spun by spiders.

Being the resourceful species we are, having plumbed various depths across the tree of life for our own material uses, it would seem a given that we would have found ways of collecting and using spider silk. We have mimicked their traps when designing fishing nets, but what about collecting actual webs from nature or extracting silk as strands we can then weave for our own uses? The history of appropriating spiders' silk is fraught with shattered ambitions, but also glistens with spectacular accomplishments.

If you were to design the perfect trap, it would be both invisible and invincible—victims would stumble into it unaware, then never pull free. Spiders are master architects of traps, and their webs succeed at achieving what seems an impossible balance of being light and delicate, yet ultra-strong. These qualities have not gone unnoticed by others. Hummingbirds will swipe spiderwebs and use them to fortify their nests. In a similar act of klep-toparasitism, humans harkening from many cultures have robbed spiders of their webs to create fishing nets and lines, for use as layering on masks, to dress wounds, or to create bags and purses.

The most astonishing use of spiderwebs originated around 1735–1825 in the Tyrolean Alps (western Austria and northern Italy), land of the Tyrolean "tongue choir" and the mummified Tyrolean Iceman, and has since been revived by only a daring few. An all-but-forgotten craft entailed gathering the fragile gossamer of a funnel-web spider (*Agelena labyrinthica* or *Tegenaria*

Coconut fiber and gummy pitch were molded to represent the human face on dance masks from Southwest Bay in Malakula Island, Vanuatu. Boar tusks jut from the corners of the mouth, and a conical framework of bamboo strips above were twirled through spiderweb after spiderweb, forming the matted, layered structure on top of the tall mask. A dense mat of spiderwebs surrounds the second mask.

domestica) or from a group of caterpillars (*Yponomeuta evonymella*), carefully cleaning and stretching their webbing, and using it as a canvas for watercolor painting or India ink fine-brush painting. If painting on a delicate web does not impress you, at least six of these were printed, requiring that a web be stretched and pressed against an etched, inked plate. Fewer than one hundred works, total, remain of "cobweb art."

The value of a spiderweb lies in its individual fibers. Could we deconstruct a web, or draw silk directly from its source? To make this a viable way to gather large quantities of silk, we would need to farm spiders and weave together threads extracted from their egg sacs or directly from their bodies. Small problem: Spiders are cannibalistic. Clever entrepreneurs seeking to achieve industrial-level harvesting of spider silk struggled with the process and were ultimately defeated by the spiders or by social circumstances beyond their control. French scientist François-Xavier Bon was the first to definitively prove that spider silk could be harvested from egg sacs by presenting a pair of spider silk gloves and stockings at a public meeting in 1709. No one attending this meeting could have imagined Bon's tribulations. First, he raised cannibalistic spiders in great numbers, housed and fed each individually, extracted and reeled silk from egg sacs, and fashioned the delicate garments. On a roll, Bon professed to have concocted a panacea for various ailments using raw spider silk that was more powerful than a silkworm moth silk potion presented at that very same meeting.

The case for spider silk was made, and it took a French entomologist, René Antoine Ferchault de Réaumur, to investigate the viability of this new enterprise as a commercial venture. Réaumur approached Bon's spider works with specific expectations in mind. Any entrepreneurial venture with the aspiration to produce silk commercially had to vie with one primary contender: How would the silk hold up to that spun by the caterpillar of the domesticated silkworm *Bombyx mori*? In 1710, Réaumur ran tests and concluded that spider silk, at least from locally collected spiders, fell far short of what would be needed to compete with the silkworm. It took twelve spiders to produce as much silk as a single caterpillar, and one pound (0.45 kg) would require the labors of 27,648 female spiders, depending on their size. These spiders would have to be housed and fed separately. Réaumur's Bon-crushing report offered some optimism, however, by recommending that tests be performed with larger spiders found in America. The satirist Jonathan Swift knew of Bon's work, and parodied it in his *Gulliver's Travels*. Swift, embellishing on Bon's spider silk-reeling practices, spins a tale of a loosely veiled Bon feeding spiders colorful flies so that "the webs would take a tincture from them," resulting in such an array of hues as to fit any client's fancy.

It became obvious that neither collecting entire webs nor reeling silk from spiders' egg sacs was going to corner the silk market. Raimondo Maria de Termeyer, eighteenth-century Spanish Jesuit missionary and amateur scientist, had a different idea: reel silk directly from a spider's spinnerets!

If you offer a spider a fly, the spider will lift the back end of their body and make a motion to spin silk around the delectable prey item. At this precise moment, you attach the ensuing silk onto a reel, and slowly turn that reel. And keep turning. Termeyer's method avoided the rough processing Bon used to treat silken egg sacs, and the resulting silk refracted light with a shimmer, sheen, and luster Bon's previous products had failed to exhibit. He overcame two additional limitations faced by Bon: Termeyer's spiders dangled from separate canes hung in his house so that the spiders could not eat each other. Instead of laboriously feeding each spider, Termeyer captured a steady supply of flies attracted to rotting meat he made available beneath the canes. Purses and stockings (for Napoleon) were made, but political upheavals seemed to interrupt or quash Termeyer at every turn. Other entrepreneurs and inventions followed, a steam engine–powered spider silk collector was employed, but efforts resulted in extravagant gifts to royalty and little more.

Though appearing suspiciously like a medieval torture device aimed at spiders, this was Raimondo Maria de Termeyer's ingenious means of extracting silk from a spider (*Alpaida latro*, pictured), as described in *Opuscoli Scientifici d'Entomologia di fisica e d'Agricoltura dell'Abate* (1807).

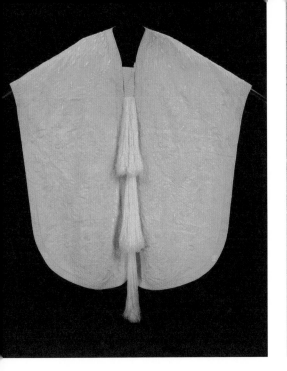

That is, until the efforts of more than one million spiders were woven into what has become likely the largest, and certainly the most elaborate, human product woven with spider silk. In a tribute to the wonder and ingenuity that sparked their predecessors, Simon Peers and Nicholas Godley took the practice of harvesting spiders' silk and weaving textiles to new heights. The silk of *Nephila madagascariensis* spiders was reeled in the style of Termeyer, and woven for thousands of hours to produce textiles that glisten with the naturally yellow sheen of the spiders' silk—a textile with traditional Madagascar motifs, and a cape adorned with delicately woven spider designs. As Peers proclaims, "The imagery it is woven with hints at the spider as creator, with echoes of myth, poetry and perhaps nightmare." Hundreds of years of trial, tribulation, creativity, ingenuity, and the unfulfilled ambitions of others culminate in this cape of echoes.

Many are drawn to the feel and look of silk, but few consider the sound of silk. Extracting, twisting, and bundling silk strands from another *Nephila*

Embroidered spider-silk cape, harvested from more than one million *Nephila madagascariensis* spiders. Simon Peers (who, along with Nicholas Godley, directed the creation of this piece) estimates one gram of silk requires collecting from 600 to 1,100 spiders.

species, Shigeyoshi Osaki has created violin strings that, he reports, produce a brilliant and unique timbre. The G string alone is composed of fifteen thousand silk filaments. If the mood struck, you could string these on a violin body made of a composite of spider silk, silkworm silk, and resin designed by Luca Alessandrini. The three strands of spider silk running the length of the violin add elasticity, and adjusting the recipe of silk and resin offers a means of sound customization new to the realm of the luthier. Eleanor Morgan, an artist who has serenaded a spider by connecting a silk strand from a spider's web to her throat and vibrating the strand by singing, also extracts silk from spiders. In an article published in the bulletin for the Royal Entomological Society, *Antenna*, she reports:

> *It is unsettling to realise that one is extracting silk from a spider and is able to feel the bodily resistance of another animal through one's hand. . . . At first I tried to work with tweezers, but the silk is more attracted to skin. So I hold and weave the silk between my hands. Often I cannot see it, so I seem to be gesturing at nothing. When collecting silk from spiders, and weaving these threads on the loom, I start dreaming of the silk, the feel of it on my hands and the look of the strands on the loom. Perhaps this is as close as we can get to the dreams of spiders.*

The use of spider silk is historically punctuated, as are so many sagas in human history, by episodes of war. War, among its list of atrocities, leaves heroes in the shadows, unrecognized and uncelebrated. There is no memorial or plaque that recognizes the efforts of wartime spiders, for instance. Black widows and other spider species were collected and kept in US spider ranches, and their silk was reeled and used to create crosshairs for gun sights and range finders during World War II. Telescopes, microscopes, surveyors' theodolites, and other optical instruments also used spider silk long before the world wars, and these uses lasted until the 1960s. Why spiders? At one-fifth the diameter of a human hair, spider silk is unmatched by any material in nature for its strength, elasticity, uniformity, and resistance to extreme temperatures. It is more homogeneous than any synthetic fiber, and can be used to manipulate light. Silk collected directly from the cellar spider, *Pholcus phalangioides*, produces a biomedically useful "super lens" that is compatible with living tissue (reducing the rate of rejection by our immune system), and, when coupled with a laser, can view the tiniest of subjects within a living body.

As we learned earlier, commercial production that relies on extracting silk directly from spiders or their webs will never compete with the scale of harvesting silk from silkworm caterpillars, and spiders are difficult to raise together because of their tenaciously solitary, predatory nature. Can we create spider silk, but remove troublesome spiders from the equation? Engineers are attempting to create a sustainable, biodegradable silk versatile enough to transform into armor, adhesive, fiber, or gel and have been watching spiders closely to learn their tricks. Spinning a strong fiber from liquid at room temperature is no trivial feat, so to approximate silk more closely means engineers emulate the natural processes involved. Biomimicry!

Dreams of spinning a synthetic spider silk have lived in the popular consciousness since at least the Silver Age of comic books. Ostracized and taunted as Midtown High's "professional wallflower," science-savvy Peter Parker attends a science demonstration that goes horribly wrong. A spider is accidentally blasted with radiation, drops onto the teenager's hand, and, in the spider's final moments, injects a potent dose of radioactive venom. Parker, dazed, exits and soon discovers that he has adopted (quite loosely) a spider's uncanny powers. A mere five panels after testing his newfound powers, he successfully spins his first wrist-activated web fluid, a key step to becoming Spider-Man! Parker's web-slinging and -swinging abilities, though a 1962 fantasy concocted by Stan Lee and Steve Ditko, clearly mimic genuine powers of spiders, and abide by James Kakalios's calculations in *The Physics of Superheroes*. If dreams of becoming Spider-Man stir within you, gentle reader, find promise in the news that replicating the qualities of spider silk appears within reach. The race to synthesize something approximating what spiders have been busy spinning for three hundred to four hundred million years has recently given birth to a hydrogel in which fibers drawn from a 98 percent water solution at ambient temperature roughly mimics natural processes observed in spiders.

There is a second path to producing spider silk that transcends mere chemical mimicry, one that invokes the mythical realm of hybrid creatures: Create something that is only part spider! That is, transforming an organism that is easier to work with than spiders into spider hybrids that serve as silk-secreting factories. Of all the organisms inhabiting Earth, which would you conscript to produce spider silk? A turtle? Pine tree? Mushroom? Hint: The mythical Chimera of Greek mythology sported a goat head on its back. The answer, according to Randy Lewis, lies in the udder of a transgenic goat.

It started when Lewis, a professor at the University of Wyoming enamored with spiders, first cloned DNA responsible for spinning silk in two marbled orb-weaver spiders. Lewis licensed the technology to Jeffrey Turner at the now-defunct Nexia Biotechnologies, and a team was able to produce spider silk in cells from a cow's udder and baby hamster's kidney. Soon after, West African dwarf goats entered the scene. The process begins with grinding up two species of spiders, tweaking a silk-spinning gene so it only switches on during lactation inside a goat's mammary gland. Next, stick the modified gene inside a goat egg, and let your first transgenic goat grow up. The goat's offspring will be genetically 1/70,000 spider, and although the spider gene will lurk in (nearly) every cell in the goat's body, it will only be active in the udder when she produces milk. Lewis moved to Utah State University with a herd of the spider goats from which he could collect the milk-silk, filter the silk, and wind it as it solidified when in contact with the air.

Introduction of transgenic spider goats marked a turning point. The silk of an arachnid could be inexpensively produced by vastly larger goats, with help from an ambitious primate species (Randy Lewis and his crew). The prospect of churning out large quantities of spider silk to be used for artificial ligaments, medical sutures, armor, and other products appeared to be on the horizon. For Turner, this accomplishment seemed only natural, given that "The silk gland of spiders and the milk gland of goats are almost identical, and teats equal spinnerets."

Other transgenic organisms, now harboring spider silk-producing genes, followed. In 2015, *Escherichia coli*—the most widely studied single-celled organism, a bacterium populating our lower intestine—produced fibers with the same toughness as natural spider silk. In 2017, *E. coli* could produce kilometer-long fibers. This was possible because the researchers carefully replicated aspects of spider physiology, including the acidic environment necessary for silk proteins to solidify into fibers. *E. coli*, although easy to rear in the lab, need to eat a special medium, so another line of research introduced a marine bacterium, *Rhodovulum sulfidophilum*, as environmentally friendly spider-silk factories in 2020. The bacterium only requires seawater, sunlight, carbon dioxide, and nitrogen. Yet another organism was involved in the first commercial production of a spider silk fabric, Bolt Threads' Microsilk—a yeast. Likewise, Lewis has spearheaded Spidey Tek, dedicated to successfully harvesting spider silk produced by alfalfa. This partial list of transgenic organisms engineered to produce spider silk has broadened our diversity of organisms from Araneae (spiders) to vertebrates (goats), bacteria (gut and marine), fungi (yeast), and plants (alfalfa). Coming full circle, our discussion of making spider silk commercially competitive with silkworms should naturally lead us to our final, inevitable recipient of spider DNA: silkworms. Indeed, Randy Lewis has collaborated with a team of researchers to investigate this transgenic approach to cultivating a low-cost source of spider silk.

It is easy to extol the virtues of spider silk, but more than two hundred thousand different animals spin silk, and each silk is endowed with its own set of properties from which we could learn and benefit. Honey bee larvae spin shorter and less refractive (shimmery) silk strands to fortify the honeycomb cells in which they develop, and Tara Sutherland genetically modified *E. coli* bacteria to produce bee silk. A Japanese lab recently explored the properties of a bagworm moth's silk (family Psychidae), with designs on its potential to be the next big renewable, strong fiber resource. Feed

silkworms and spiders carbon nanotubes and they will spin reinforced silk. Apply technologies such as 3D-printing and silks may one day surpass anything natural silks have to offer.

The Silk Road transcended its role as a mere avenue by which commercial goods were transported. It endures to facilitate cultural exchange, stimulate economies, and prod human ingenuity. Though silk continues to bind us together, our global ambitions are changing. As we grow beyond our means, the most promising future manifestation of the Silk Road may lie in our desire to realize environmental sustainability, a road that might allow us to learn from but diminish the harm we inflict on our silk-spinning relatives.

Wax

Your new home is a hole in a tree. It's spacious, offers some protection from the elements and invaders, but it lacks infrastructure. Where will you store your food? How will your youngest family members stay warm? Where will everyone sleep? It won't be long before this vacant, hollow shelter is bustling with tens of thousands of your siblings. Their needs will be many, and the first step will be to build a home that can carry you through winter. As a western honey bee, you are a cavity dweller with an uncanny ability to create your own building material. Paper wasps build their nests by mixing chewed wood with saliva and laying row after row of delicate paper pulp. Mud dauber wasps gather moist soil and mix it with saliva. Some termites build their nests from their own fecal droplets. But you, like other honey bee species, can exude translucent white flakes of wax from your abdomen. These soft, pliable plates can be chewed, sculpted, and reshaped to form a honeycomb that is both lightweight and durable, with cells that can store food and developing brood, serve as places to warm these juvenile bees, or to escape the hubbub and sleep. The same honeycomb serves as a stage on which a nestmate can dance to advertise the location of food or water she has discovered in a far-off land.

This comb doesn't come without a cost. For you to become a productive wax worker, your early days as an adult better have included eating pollen to help your body develop fat cells, and your time now must be spent gorging on honey. Pollen and honey are hard-earned ingredients, requiring many potentially risky trips made by your nestmates to flowers. The honey you consume is metabolized in fat cells associated with eight wax glands in your abdomen. Wax glands then convert the honey into thin layers of beeswax

Honeycomb hangs
in parallel from a
ceiling, constructed
from wax scales
exuded by western
honey bees (*Apis
mellifera*).

that accrue into plates. For over thirty million years, honey bees have secreted wax to build comb, and your ancestors' architectural legacy lives on, concealed within the hollow cavity of your arboreal home.

Though perfectly suited to being a comb-building material, your wax is a versatile plastic and its discovery by a resourceful and visionary primate was inevitable. Early humans likely discovered the utility of beeswax after a chance tasting of its contents. A partially scavenged honeycomb might have given hunter-gatherers their first mouthful of honey, as well as of protein-rich bee brood. Honey became a sought-after treasure, but the wax comb containing it—being sticky, pliable, water resistant, and long lasting—became indispensable. Beeswax can be used as an adhesive, lubricant, varnish, polish, and in cosmetics and foods. As a coating, beeswax keeps water off or keeps water in. Beeswax wrap is a food wrap material used instead of single-use plastics to keep food fresh. Beeswax can control bleeding from bone surfaces, and is used as a way of slowly releasing drugs or bait poisons. As beeswax candles burn, cathedrals are brightened and prayer intentions are symbolized. Pigmented and melted, beeswax is the traditional medium in encaustic painting. Beeswax never goes bad, so evidence of its use pops up on Viking ships, and among Roman and Germanic tribal remains. Egyptians sealed the noses, eyes, and mouths of their mummies with beeswax, and used it medically, cosmetically, for rituals, and to waterproof boats. The head of Nefertiti, one of the most recognizable pieces of art in the world, is a mixed media work, with beeswax binding the charcoal black pigment.

Beekeeping—the practice of providing an artificial home for bees—was well established in Egypt by 2450 BCE. A relief in the sun temple of pharaoh Nyuserre Ini (aka Newoserre Any) clearly documents beekeeping practices, as do later reliefs and paintings. Beeswax and honey have been recovered from ancient Egyptian tombs. Even the smell of bee products can be detected and analyzed from ancient remains, as Italian chemist Jacopo

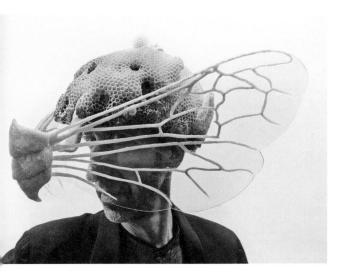

Two of my alter egos, with garb from a clandestine armory of hive helmets. Placing a recycled bicycle helmet into one of my honey bee hives, workers build comb until I extract the helmet and modify. The winged helmet includes Plexiglas wings, beeswax, and insect pins (for Hymenoptera-defining hooks called hamuli, not pictured in this early version).

Detail of a second hive helmet sculpted with wax foundation (beeswax sheets, impressed with a uniform base of honeycomb cells, used by beekeepers to elicit comb building).

La Nasa and colleagues report in the romantically titled study, "Archaeology of the invisible: The scent of Kha and Merit."

These pictures and odors offer clues about advanced beekeeping practices in ancient Egypt, but studying actual beehives would help answer unresolved questions about Egypt's methods and approaches to beekeeping. No hives have ever been discovered among the temples or city ruins. The oldest known beehives were not discovered in Egypt, but, rather, in Tel Rehov, Israel, in 2007 by Amihai Mazar, archeologist from the Hebrew University of Jerusalem. Having survived the ravages of time and fire, hives built of clay and straw circa 900 BCE offer charred and fragile remnants of bees, and evidence about the human civilization that kept them. For instance, Mazar suspects that beeswax, not honey, may have been this apiary's most important commodity. With one hundred active hives in its heyday, the apiary may have supported a local copper industry at the time.

Copper objects were cast using the lost wax process, in which metal is used to replace a delicate wax prototype. Objects are first sculpted in wax, the wax is encased in a mold, the mold is heated to melt out the wax, and

molten metal is poured in to fill the space left by the expunged wax. When cooled, the metal forms a sturdier replica of the original wax form. This was a common practice around the world and dates to at least the third millennium BCE in the Middle East. Troves of metal material culture owe their existence to bees and their wax. Lost wax casting is responsible for everything from life-size bronze sculptures of humans (Benvenuto Cellini's *Perseus with the Head of Medusa*) to tiny brass figures by the Ashanti of West Africa. The vulture and cobra adorning Tutankhamun's funerary mask, symbolizing his rule over Upper and Lower Egypt from 1334 to 1325 BCE, were made by the lost wax process. Stingless bees' nests were available in the Americas, and the Inca cast delicate filigree jewelry, replacing stingless bees' wax with gold. Though not cast in metal, Aboriginal Australians used the wax and resin from stingless bees' nests to paint and sculpt objects of magic and sorcery.

The oldest evidence of beekeeping (circa 2450 BCE), depicted as a relief from the great solar temple Shesepibre, under the reign of Nyuserre Ini (circa 2474–2444 BCE). The relief, on five broken limestone blocks (two of which still retain their color), has been reconstructed and is now displayed in two sections at the Neues Museum in Berlin.

TOP: Beekeepers, depicted as a relief in the tomb of Pabasa (tomb TT279), during the twenty-sixth dynasty of Egypt (664–525 BCE). One man pours honey (*above*), another appears to be in a pose of praise or worship, and honey bees, shown in their hieroglyph form, fly to either side of a stack of oblong beehives.

Honey bees, symbols used for everything from breakfast cereals to religions, adorn the Barberini shield at the Palazzo Barberini in Rome, flanked by a pair of sphinxes. Cardinal Maffeo Barberini was elected pope in 1623, taking the name Urban VIII.

Long, long before any of these reliefs, ruins, and relics were shiny and new, the origin of humans' relationship with honey bees was recorded in ocher on limestone. An anonymous painter depicted the robbing of a honey bee nest on the wall of a Spanish cave. The painting shows a rope ladder tied into loops, which was attached to the top of a cliff, and supported by a pole against the cliff face. Ascending from the bottom, a thief bravely weathered a whirling barrage of bees. Despite the threat of thousands of venom-injecting, barbed stingers, honey bee nests were a target for human foragers. The practice of robbing honey bees' nests became a fixture in Levantine culture, and this cave wall and ceiling scene is the clearest surviving depiction. The daring heist is captured 7,500 years before sunlight passed through a tiny hole and altered light-sensitive substances to form the first photographic image. Discovered only recently, the painting has outlasted the life of the honeycomb robber, the artist, and three hundred generations of their descendants. Humans are painted robbing other honey bee nests in Spain, as well as in India and, most

A sophisticated act of honey robbing using a looped rope ladder, documented in paint about 7,500 years ago on the ceiling and wall of a newly discovered cave site in Barranco Gómez in eastern Spain

0   20 cm

commonly, Africa. Even in Australia, where honey bees did not exist and the honey of stingless bees was the local prize, a cave painting of a possible honey robber stands by two stingless bees' nests. The figure is holding a stick, like honey robbers in Australia do today.

Caves, as shelters from the elements, were attractive havens for cave artists, as well as for nesting insects. The pairing may seem random, but insect nests offer clues that have helped shed light on one of the great perennial secrets concealed within most cave art—their age of creation. Archeologists, faced with limited means of estimating a cave painting's age, have discovered ways to use wasp nests as a guide. Mud dauber wasps build nests of clay, and these can be irritating fixtures for homeowners and art conservationists because the nests are difficult to remove from the surfaces to which they were cemented. To be a successful mother, a mud dauber wasp collects moist soil during a wet season, finds a shelter from the rain, and plasters the mud against a surface so the resulting nest bearing her offspring will not fall and fail. She does such a good job that remnants of her nest can last tens of thousands of years. We know this because organic material in her nest degrades at a predictable rate, and you can compare specific carbon-rich substances in an old nest with the same in a new nest. Once you have an estimate for the age of the nest, look for its position relative to the painting. If a cave painting lies beneath a mud dauber's nest, the painting must predate the nest, but if the paint covers a nest, the painting must be younger than the nest. In this way, a wasp's ancient architecture can establish the minimum or maximum age of a human's painting. Mud once seen as marring spectacular rock art can now be sampled, processed, and prized as data-rich evidence that can tell us about the history of insects and of humans.

Cave art uniquely transports us in time, allowing us glimpses of early human practices through an artist's eyes. By comparatively studying these interpretations with evidence of more recent or present-day practices, archeologists and anthropologists can expose what elements of a tradition have been maintained, or hypothesize how a tradition has changed. The honey-robbing scenes captured in ocher help us connect the actions of ancestors to nest-hunting cultures today.

There is more than one way to rob a bee nest of its valuable contents, and the allure of honey, wax, or the bees themselves has inspired different honey-hunting strategies across the world. Travel to Niassa Special Reserve (Mozambique) or the Lake Eyasi region (Tanzania) and you will find one honey-hunting tradition so unlike others that to appreciate it demands more than observing humans robbing bees' nests. It demands talking to birds.

When we name species, every now and then the name indicates some aspect of the species' behavior, and nowhere is this clearer than in the bird species name *Indicator indicator*, known commonly as the greater honeyguide. This bird, when eager to eat its next meal of wax, makes a chattering call to attract human honey hunters. If you, as a prospective honey hunter, were to follow, the greater honeyguide would take you from tree to tree, indicating the direction to a honey bee nest. Once there, the bird stops and sits quietly and waits patiently. The polite expectation is for you to open the nest, usually using smoke to calm the bees, and forage, leaving a gift of wax comb

for the bird as encouragement to repeat the mutually beneficial exchange in the future. If you are a member of the Yao people of Mozambique, you may not wait for a honeyguide to initiate a conversation, but begin the dialogue yourself by making a loud trill that ends with a short grunt. This tells any greater honeyguide in the area that you are ready to begin the process of jointly traveling to a nest.

Local honeyguides have learned to associate the sounds of your specific culture, so if your sounds are just right—whether they be words, trills, grunts, or whistled melodies—you increase your chances of engaging a willing honeyguide twofold (33 to 66 percent), which more than triples your chance of finding a bee's nest than if you had attempted the venture without your feathered guide (17 to 54 percent). Behold the most remarkable relationship humans have ever fostered with wild birds—a tradition that may span the entirety of human history. From the bees' perspective, this alliance between primate and bird represents a persistent threat, and this threat underscores the value of their wax.

Beyond the Bees

Travel to one of the oldest public science museums—La Specola in Florence, Italy—and you will be greeted with scenes of syphilis, the plague, and a decapitated head, all masterfully sculpted in a beeswax mixture by Gaetano Giulio Zumbo at the end of the seventeenth century. A room of anatomical waxes by his successors offers oddly timeless, perfect replications of organs, embryos, limbs, and bodies. Sculptors used the translucency of wax

FAR LEFT: Seliano Rucunua is briefly holding a wild-caught male honeyguide (*Indicator indicator*) during research.

Kaliambwela is climbing a tree to access a honey bee nest found by a honeyguide.

THIS PAGE: A female honeyguide tends to the spoils of a honey-hunting event, left for her in an act of reciprocation by the person who opened the nest and extracted the comb.

to capture the subtle colors and textures of the human form. Alluring and macabre, full figures appear unsettlingly alive after two hundred years of voluptuously lounging, encased, on silk sheets, and propped on moth-eaten velvet cushions. These sculptures, one of which can be dissected to reveal layers of viscera, acted as surrogates for cadavers to educate doctors in training. But as an instructional model, "Each is far more lifelike and lovelier than seems strictly educationally necessary," observes Joanna Ebenstein, author and director of the Morbid Anatomy Museum in New York City.

When something fabricated resembles humans closely, but the resemblance is disquietingly imperfect, we find these fabrications to fall within an "uncanny valley." The wax filling the uncanny valley populated by these medical models includes "Chinese wax," produced by scale insects (*Ceroplastes ceriferus* and *Ericerus pela*) that scarcely resemble insects at all. An adult female is a sessile, plant-sucking, wax-exuding blob. Concealed within her small mound of sticky white wax, she is wingless, soft-bodied, and lacking the obvious hallmarks of an insect. Strain closely with magnification and you can find the minuscule vestiges of legs, antennae, mouthparts, and simple eyes on her underside. Used in China as a medicine, polishing agent, and for candles and temple images, her wax is hard, white, and melts at higher temperatures than beeswax. Though far more expensive and more difficult to obtain, this wax has unique, desirable attributes.

Waxes are exuded by a wide variety of insects, with a scant few waxes assessed for their potentially unique, beneficial properties. Some secrete so much wax that you might misidentify them as wind-dispersed seeds, or

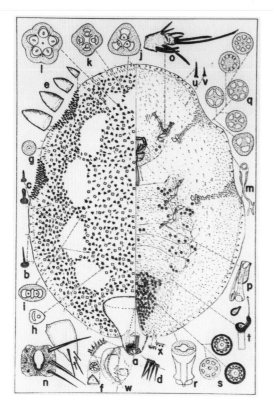

white fungal growths. Closer inspection reveals insects sporting waxen trails of flossy tails, fluffy jackets, or glistening scales. Insects often enthusiastically exude copious quantities to conceal and protect themselves, or, as we've seen, to build an all-purpose comb to support a family of thousands.

The most basic use of wax for insects is to waterproof their bodies. They exude a thin coating that covers their exoskeletons. If in a society, insects' waxy coating is also used to chemically recognize each other, to distinguish enemy from ally. Forensic entomologists are attempting to use these same chemicals found on immature flies to pinpoint the time of a person's death when solving murder mysteries. A standard approach to determining time of death is to identify the developmental stage of insects on a corpse and to calculate how long it took the insects to reach these stages, taking into account environmental conditions like temperature. As immature flies age, the chemical signatures on their bodies change, so analyzing their surface wax may make the process of reconstructing a crime scene easier.

Even ancient human remains can be dated using an insect bodily product—though in this case the product is not a wax, but a glue used to adhere eggs to our hairs. Body and hair lice are so specialized that they are not long for this world if separated from our bodies, and one facet of this intimate relationship involves cementing their eggs on the shafts of our hair. If you have ever tried to "nitpick" with a louse comb, or attempted other desperate measures to remove louse eggs, you are keenly aware of their resilient properties. Adhered so stubbornly to our hair, louse eggs can be seen on ten thousand–year-old remains. Surprisingly, most such remains are speckled with nits. Even more surprisingly, invertebrate biologist Alejandra Perotti led a team of scientists to rescue ancient human and environmental DNA trapped within louse cement to learn about individuals' lives two thousand years ago. Peoples' dispersal patterns, health conditions, and cause of death can emerge from DNA evidence, though extracting DNA without damaging human remains can be a challenge. As Perotti and colleagues have revealed, human lives otherwise lost to time can be respectfully revived and appreciated one hundred generations later, thanks in part to even the most despised of insect parasites.

Nits of head lice (*Pediculus humanus capitis*) attached to hair shafts from the head of pre-Columbian human remains (*left*), and nits and adult lice caught between the teeth of a comb used to remove them (*right*)

 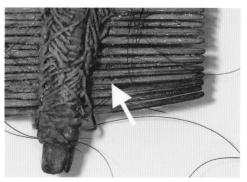

Honey

Plants perform two tricks that have given much of the rest of life an opportunity to exist. First, plants consume light, basking in the sun's rays and converting the photons of light into usable forms of energy. Second, plants can "fix" carbon, which means they inhale carbon dioxide, then take this gaseous waste product of ours and extract its carbon atoms to use as building blocks. Using the energy converted from sunlight and the carbon from carbon dioxide, plants construct sugars. In flowering plants, some of this sugar is invested in nectar, the sweet, mildly nutritious beverage used to entice pollinators, like honey bees, to visit flower after flower, transporting pollen in the process.

An experienced honey bee can anticipate when a flower will bloom, and beat her competition by arriving early to lap up the sweet beverage it produces. Even before returning to the nest, the nectar she stores in her honey stomach is already partly digested and turns into honey. One enzyme in her body converts nectar's sugar, sucrose, into the simpler sugars glucose and fructose. A second enzyme converts some of the glucose into hydrogen peroxide and gluconic acid, both of which make honey inhospitable to bacteria and fungi.

As a protective bonus, the high volume of sugar in honey draws water out of the air, or out of the bodies of invaders that would otherwise spoil the honey. If too much water is in the mix, honey spoils or ferments (think of the alcoholic beverage mead), so worker bees fan their wings to assist in the evaporation of water from honey until it reaches just the right concentration. At this point, bees cap the cells with wax and store the viscous liquid until needed. This conversion of nectar into honey makes it possible for honey bees to store food for extended dry spells. No colony of honey bees can overwinter in a cold environment without ample supplies of honey, and life far from the equator can mean long, cold winters.

So effective is a honey bee's conversion of ephemeral nectar into perennial honey that honey has been used as a preservative. Damaged by a storm in the Bay of Bengal, an English ship carrying Robert Knox and his father's crew landed in Ceylon (now Sri Lanka) in 1659. Enticed with gifts and invited ashore by Rajasinghe II, the third king of the kingdom of Kandy, the crew was taken captive. Held for nineteen years until his escape, Knox was free to farm, knit garments, and work as a money lender. Following his escape, Knox documented the local practices of the "wild" Vedda people of Ceylon in what became a famous travel log. In it, he made note of the use of honey to preserve meat: "They cut a hollow Tree and put honey in it, and then fill it up with flesh, and stop it up with clay. Which lyes for a reserve to eat in time of want."

One thousand years before Tutankhamun was entombed with a jar of honey to sweeten his passage to the afterlife, the Kura-Araxes culture in the Caucasus buried a pantry full of foods and weavings coated in honey. Alongside this 4,300-year-old honey-preserved cache were human remains. Bones speckled with pollen and bee bits showed signs of having been prepared for a honey-soaked journey to the afterlife. Ancient Assyrians also embalmed with honey, and the bodies of Alexander the Great (323 BCE) and Eastern Roman Emperor Justin II (578 AD) may have been submerged in honey-filled caskets. Then came the earls. Sealed in coffins of lead, buried in a vault below a monument lie the Earls of Southampton, immersed in honey. Their honey-drenched eternal resting conditions had been documented, but when one casket was accidentally damaged more than one hundred years ago, a worker felt compelled to verify the story. Liquid seeped out. It was obvious what the worker needed to do to satisfy his hunger for truth. All it took was a swipe of the finger and a small taste from a leaky seam.

Some honey tales are better documented than others. Pliny the Elder, author of the largest single work to survive from the Roman Empire—*Naturalis Historia*—wrote a chapter called "Marvellous Births." Among records of humans giving birth to an elephant and a serpent, Pliny claims to have seen a hippocentaur (half human, half horse) preserved in honey, brought from Egypt to the emperor Claudius Caesar. Another tale of honey preservation lurking in the literature is a secondhand account of the Arabian "mellified" man. Elderly volunteers were said to spend their final weeks of life consuming only honey, bathing in honey, eventually excreting only honey. After death, their bodies were interred for a century in a casket full of honey. Once exhumed, the mellified men were sold as human healing confections, a tiny piece being sufficient to heal a broken bone. The lore of the mellified man was reported with due skepticism by Li Shizhen in a sixteenth-century Chinese volume of herbology "for the consideration of the learned."

Honey preserves, but can it heal? The mellified man is not the only questionable case of honey as curative. First, suspend your disbelief when reading the following bogus biology from a book by Bernard Read collecting historical Chinese medicines: "All bees in making honey carry human feces to the flowers to ferment the nectar and so produce a ripe product, just as one uses yeast in the making of malted rice candy." If you were able to swallow that, you can now marvel at honey's healing properties reported by Pen Ching in the same volume:

> for indispositions of the chest and abdomen, all types of convulsions;
> it quietens the viscera; for all kinds of weakness, a vitalizer, a digestive
> tonic, a sedative, an antidote for poisons; it can get rid of numerous
> sicknesses. It is compatible with all medicines, taken for a long time it
> strengthens the will, lightens the body, and keeps away hunger and
> senility. It confers everlasting life and one becomes a divine immortal.

Could there be a more comprehensive elixir? Either too few people have taken Pen Ching's findings to heart, or there are honey-devouring immortals among us. Everlasting life aside, honey does have inherent qualities that speak to its medicinal value. Natural selection would spell doom for honey bees storing contaminated honey, so the antiseptic properties of honey have factored strongly in the evolutionary success of honey bees. Honey staves off invaders with the three-punch solution of being acidic, containing little

moisture, and also hydrogen peroxide producing damaging free radicals. Hydrogen peroxide is produced when honey is immersed in recently dead animal tissue or when diluted by body fluids, which could help explain why honey has the power to preserve tissues or heal wounds.

The Beekeepers (circa 1568), by Pieter Bruegel the Elder

Although honey is mostly sugar and water, and its makeup depends on the flowers from which bees gathered their nectar, honey contains roughly two hundred substances, including vitamins, minerals, amino acids, and enzymes. Ancient Sumerians, Egyptians, Assyrians, Chinese, Greeks, and Romans applied honey to wounds or to remedy gastrointestinal issues. Sumerians recorded the earliest known prescription using honey in cuneiform between 2100 and 2000 BCE. To the Egyptians, honey was the most versatile of remedies; nine hundred treatments for ailments are listed in medical papyri, and about five hundred of these include honey. Sometimes honey was intended as a binder, or just to make the medicine tolerable to the taste, but many treatments prescribed honey as an active ingredient. As an illustration of Egyptians' liberal medicinal use of honey, they prescribed it to induce birth, or, when combined with natron, sour milk, and crocodile dung, to prevent getting pregnant in the first place. The great Greek physician Hippocrates also prescribed honey for contraception, as well as for baldness, healing wounds, and a variety of other purposes. Modern techniques reveal that honey can inhibit about sixty species of bacteria, as well as some fungi and viruses. Evidence is growing to support many of the basic medicinal uses of old, and honey may play a larger, not lesser, role in our medicines of the future.

Honey may also serve as a future indicator of air quality, offering sweet warnings of threatening lead levels in our environment. For each beehive, thousands of foraging honey bees collect nectar from their surroundings, bringing back a chemical signature of their habitat. Monitoring the honey from different locations might give us a real sense of where contaminants exist and what is producing them. Honey, once again, for our health.

There may be one additional property of honey worth mentioning. It is tasty. Drip lemon juice onto your tongue and your brain processes the stimulus as being sour. Do the same with something salty, bitter, sweet, or savory/umami, and you will perceive the suite of our taste sensations. Tasting something bitter can alert us to possible poisons. Tasting something sweet can direct us to high-energy foods. Cats don't need or have the ability to taste sweets, but honey bees and humans do. Our palates are so attuned to sweet foods that we have preferences for honeys depending on the sweet sources honey bees collect to produce them. Is your preference for a light, neutral, floral honey, or a dark, thick, musky honey? Temperate honey bees can collect from plants like acacia early in the season or buckwheat later in the season. When the last blooms have withered in autumn, honey bees need not close up shop. Some can turn to a "honey-like dew" that Pliny the Elder, our go-to Roman scholar, speculated originated from the heavens:

> Whether it is that this liquid is the sweat of the heavens, or whether a saliva emanating from the stars, or a juice exuding from the air while purifying itself, would that it had been, when it comes to us, pure, limpid,

and genuine, as it was, when first it took its downward descent. But as it is, falling from so vast a height, attracting corruption in its passage, and tainted by the exhalations of the earth as it meets them, sucked, too, as it is from off the trees and the herbage of the fields, and accumulated in the stomachs of the bees—for they cast it up again through the mouth—deteriorated besides by the juices of flowers, and then steeped within the hives and subjected to such repeated changes—still, in spite of all this, it affords us by its flavour a most exquisite pleasure, the result, no doubt, of its æthereal nature and origin.

Honeydew is actually the liquid waste of bugs feeding on plants—specifically, some scale insects and aphids. Aphids need to suck a lot of sweet sap to acquire trace quantities of protein, so a great deal of what they imbibe goes right out the other end as they feed. Other insects take advantage of this sweet residue, including honey bees. Although the origins of honeydew are not as cosmic as Pliny imagined, he astutely considered the factors affecting the substance's transformation, from plant to bee stomach to bee nest to human mouth. In this case, the exquisite pleasure we perceive comes from a bee regurgitating the excreta of true bugs, not the saliva emanating from stars.

Countless poems and songs wax eloquent with honey metaphors. To be someone's honey is to be a source of sweet enjoyment. But its sweetness holds greater riches—of the caloric variety—and honey may have given our early ancestors an edge. Emphasis is always placed on the legacy of our ancestors hunting animals and gathering plants, but robbing bee nests is also an ancient human pursuit. Robbing nests meant devising strategies of finding and acquiring nests and innovatively using tools. Could it be that such innovations and supplementing diets with bee nests' energy-rich contents allowed our human ancestors to rise above their competition, and fueled their ever-expanding brains?

Lacquer

et me paint a picture of you, enjoying music while relaxing in a room. You are sitting in an intricately carved chair, rocking back and forth to the rhythm of music from a bygone era. Your garments are comfortable and colorful, hair is perfectly in place, oil paintings and textiles on the walls. Wood furniture, trim, and floors glimmer with a waxen sheen. Bowls of shiny fruits and candies are within arm's reach. Every object around you is composed of, or covered by, bug secretions.

This final detail may or may not add to your sense of repose I took such great pains to cultivate. You may be wondering what insidious insect has infiltrated your imaginary nook and spread its innards outward. What steps are necessary to extract this foul bug-stuff from your safe haven? What anomalous nightmare have I situated you in? The bug is the lac insect (*Kerria lacca* or close relative), and lac insects secrete lac, which humans process into shellac. Though portions of this scene were historically more common than they are today, no element of this picture is anomalous. The wood furniture, floors, and trim, the food bowls, oil paintings, and the shiny fruits and candies could all be coated with shellac. The plaster walls might be sealed with shellac, your nails brushed, shoes polished, and your perfect coiffure sprayed with a thin coating of the same. The old 78-rpm record is primarily composed of shellac as a binder. A dye made from the hemolymph (blood analog) of lac insects could be coloring your paintings and textiles, the clothes on your body, and serving as a cosmetic to tint your skin. There is almost no surface that couldn't be covered by the exudate of lac insects.

A Sanskrit tale from the Mahabharata tells of a palace made of shellac. A blind king, Dhritarashtra, stewed with envy as he learned about the conquering successes of five powerful warriors, the Pandava brothers. A plot is devised to undo them. Dhritarashtra's son Duryodhana ordered the architect Purochana to build a palace of shellac, ghee, and other combustible materials. Tricked into staying in the palace as guests, the Pandava brothers were saved by an uncle's warning, and fled through an underground tunnel as the palace was set ablaze. Purochana perished in the flames meant for the brothers, and any evidence of a house of lac was incinerated.

Why do lac insects even create the stuff? Lac is a lac insect's shelter through which the insect can move and feed on sugary plant phloem. The scale insect is a soft, vulnerable being, and protection is paramount. Exude a liquid that hardens and you have a refuge from the dangers lurking outside—that is, unless humans discover your secret powers of protection. Some enterprising person in ancient times saw the value in an insect's defense and learned to process the barrier as a protect-all coating. It isn't too great a leap to imagine someone observing

According to museum records, lac from Mexican lac scale insects (*Tachardiella mexicana*) was gathered from branches, rolled into cylinders, and consumed as a delicacy by the Rarámuri (Tarahumara) people of Mexico. Lac could be gathered in summer, then boiled in water and eaten as a sauce during winter. "It is refreshing and has medicinal properties," noted the author of the museum catalog card over a century ago.

the durable material on tree branches and scaling up the operation hundreds of thousands of times to coat a piece of armor or a bowl, or to stain fabric.

The value of a liquid-turned-durable-solid finish was recognized across human communities and cultures. A world away, members of the Maya civilization gathered related insects (*Llaveia axin*) from which they produced fine, durable finishes, but the source of this lacquer was fat from within female insects' bodies and not from the hardened globs of shellac produced by their lac insect relatives. Today we produce synthetic lacquers and plant-based lacquers, but the love of the lac insect continues after three thousand years, and an industry that processes the hardened insect secretions on sticks still employs millions of people in India and Thailand. It remains the only commercial resin of animal origin.

If you are willing to settle back into the rocking chair and listen to the 78-rpm record in our imaginary room of repose, there is a lesson to be learned. Until I alerted you to the presence of bug secretions coating everything in sight, you would not have noticed or imagined it. If we are oblivious to the fact that a single insect species could coat our room and our body, what other insect products are hidden in plain sight?

Color

Colors affect the way we feel and the way we think. Advertisers, architects, artists, shop owners, and therapists all carefully select colors to elicit specific behavioral outcomes. Moods are managed by lighting technicians. Emotions are evoked by fashion designers. Temperaments are tweaked by interior decorators. Most of us are primarily visually oriented, and colors are important vehicles for guiding our conduct.

Just how important are colors to us, and what happens if we choose the wrong hue? If you want to remain calm, for example, would you choose to sit in a red room or a blue room? Stranded in the woods, if your life depends on you eating an insect, would you gobble one that is camouflaged or one that is bright and colorful? Given a choice of uniforms to wear during a competition, which color uniform would you choose?

Before we address any of your choices to these probing queries, let us first consider what colors we associate with power, status, and hierarchy. Picture in your mind, right now, members of royalty, nobility, or religious authority. What are they wearing? Do specific colors complement the characters in your imagination? If red or purple prevailed, there is historic precedent to your regal vision. Emperors and kings have worn red or purple since ancient times. Purple meant power to Phoenicians and rulers of the Byzantine and Holy Roman Empires. The Caesars adopted purple as their official color, and if any non-Caesar were caught wearing a purple garment in Rome, such insubordination could mean a death sentence. Imperial status in Europe switched from purple to red when Constantinople fell to the Ottoman Turks in 1204. Hundreds of years later, Catholic cardinals once cloaked in purple switched to red, though bishops still don purple. In Japan, purple is associated with the emperor, whereas red signifies communism the world over, and conservatism in the United States. What could possibly be so special about these two colors?

For starters, red is visually striking. Reds fall on one extreme end of the narrow band of electromagnetic energy visible to humans, and unless printed using the CMYK (cyan, magenta, yellow, black) color scheme, purple is a mixture of red with blue. In nature, red often serves as a warning to predators of something toxic or venomous. Red insects (ladybird beetles, milkweed beetles, milkweed bugs, burnet moths, velvet ants), fish (scorpionfish), amphibians (strawberry poison-dart frogs), and reptiles (coral snakes) often harbor toxins. Think of the stark red markings of black widow and redback spiders. Some would-be predators instinctively avoid—or learn to avoid—such warnings, or suffer the consequences. We humans are particularly responsive to the color red. Depending on context, seeing red can bias our attention, influence how sexually attractive we find someone, or make us respond faster and with greater force.

Since red is associated with desire and heightened anger, scientists wondered if wearing red could also increase an athlete's chance of winning a match. Well, if men's boxing, tae kwon do, Greco-Roman wrestling, or freestyle wrestling at the 2004 Olympics Games are a guide, then red triumphs over blue. Russell Hill and Robert Barton published a study of these competitions and made bold associations between the evolution of what goes into selecting mates (sexual selection) and how wearing red can enhance an athlete's performance. If true on wrestling mats, is such a bias also true on the battlefield? Did sporting red coats in any way clinch British victories in the battles of Waterloo, Salamanca, or Inkerman? Then again, if seeing red enhances performance, wouldn't the opponents *not* wearing red be expected to prevail, considering that they are seeing their red-clad opponents?

It might appear that red is a good choice if you wish to woo a mate or battle an adversary, but reserving red or purple for history's elite may have had little to do with the physics or inherent qualities of color, or with psychological associations related to sexual desire or domination, though these qualities probably didn't hurt. The single most important factor for selecting a color to signify power was availability of the color. If a commodity is limited, one can monopolize it. Dominating a limited resource, both in nature and in human society, is a path to power. The easiest way to display one's status and wealth is to wear something symbolic of status and wealth. If what you wear is difficult to obtain and, consequently, prohibitively expensive, you are displaying an honest indicator of your material worth. Red and purple are both difficult to come by in nature, and before synthetic versions of both became available, harnessing vibrant, long-lasting versions of these colors was a challenge.

Take flamingos, for example. If a flamingo wishes to impress a mate, the very color of their plumage depends on the crustaceans they've eaten. The crustaceans pass along carotenoids—the class of pigments that give carrots their orange and salmon their red colors. The more carotenoid-rich food a flamingo eats, the redder their plumage becomes. In an environment with limited carotenoids, redder flamingos are the star foragers. Since carotenoids stimulate the immune system and appear to positively affect our health, flamingos may benefit from the same, so displaying evidence of more carotenoids may indicate greater health, as well as a greater ability to access a limited resource. The honest indicator of their effort fades with time, however. The solution to update their status is to spread an oil from a gland at the base of their tail over their feathers. Greater flamingos spread this oil when preening, and the oil contains cosmetic pigments that the flamingos, once again, obtain from their meals. The only way any animal with a backbone can acquire carotenoids is from the environment, and flamingos, scarlet ibises, and roseate spoonbills need to gobble crustaceans to display their reds, just as male American goldfinches and yellow warblers eat carotenoid-rich plant material to display their yellows. Though it can take work to acquire and maintain reds and yellows, these hues are far more common than the elusive purple, a rarity in nature.

The prized source of purple that was responsible for much of this color adorning the ancient elite mentioned previously came from several species of predatory snails (*Bolinus brandaris*, *Stramonita haemastoma*, and *Hexaplex trunculus*, belonging to the family Muricidae). Knock one of these snails with a hammer to expose its hypobranchial gland, cut the gland, and a defensive mucous seeps out. When exposed to light and air for a specific length

Not your typical view of cochineal bugs: Two adult males, bearing wings, long antennae, and visible legs, and wispy white wax tails originating from a pair of tail-forming pore clusters. In contrast, the red glob above the central male is an adult female. Note the hint of yellow lurking behind her (*right*). This is a scale-feeding snout moth caterpillar (*Laetilia coccidivora*), a predator that uses the carminic acid of their cochineal prey to defend against their own predators.

of time, you have the dye we know as Tyrian purple. Though the snails can be "milked" (prodded to defensively exude the mucous), the dye was destructively harvested, with roughly ten thousand snails killed to produce one gram—enough to dye a single garment, or at least its hem. The labor and time to produce such a tiny quantity makes it clear why Tyrian purple was worth more than its weight in gold and became the preeminent color of royalty.

Purple came from mollusks, but red came from . . . a worm-like, snail, berry-seed (more on this dubious identification shortly). Colonialists sweeping through civilizations in the Americas discovered plants, animals, and cultural practices wholly new to the Old World. Surprise after surprise came back on ships and by word of mouth, and some of these discoveries garnered great wealth for European conquerors. A key commodity for the Spanish Empire was a dye so potently red that its like had never been seen before.

A more typical view of cochineal insects—cotton mats of white wax on a pad of *Opuntia* prickly pear cactus, with females hidden below

But what was new for those on one side of the Atlantic Ocean had been used for over one thousand years on the other side. The Aztec and Maya civilizations each harvested the dye and used it to color fabrics a range of hues, including a brilliant scarlet. Aztec Emperor Motecuhzoma Xocoyotzin, worshiped as a god by his people, was paid tributes, including bags of what became known as "cochineal" ("coccinus" is the Latin word for scarlet-colored). When Spanish conquistadores sent tons of cochineal to Spain, a rapid fervor overtook Europe and this sensational new red was used to dye textiles, artists' palettes, and became nature's hot new product. Cochineal was second only to silver in terms of Spain's New World export.

The world clamored over cochineal, having no idea what it was. Did it come from a plant or animal? The debate raged for centuries as to whether the dark, dry pellets or prepared paddies of cochineal were berries, seeds, fruits, snails, or worms. At the time, maggots and bees were thought to spontaneously appear from rotten meat, so cochineal could very well have originated from a plant-turned-animal, or "wormberry." Infiltrators and spies from England, France, and Sweden attempted to steal the secret from

Mexico and dismantle Spain's monopoly on the enigmatic dye. Where to find it? How to recognize it? The closest success came from one determined French botanist who, at last, was able to see cochineal in the wild before smuggling some out. But it wasn't red.

When you first see cochineal, it is concealed in cottony white wax clumped on the pads of prickly pear cacti. Remove some of the white wax and you can find scale insects, like those producing Chinese wax, busy sucking the plant's sap. When exposed, the bugs don't exhibit any dramatic evasive maneuvering. Their soft, plump bodies stay put. If you flip one over and use a good magnifier, which was not common at the time, you can make out tiny legs and wee mouthparts. They do not have clearly segmented bodies or wings, like your typical insect. Only when a minuscule male flies by to find one of these sessile females can hints of their true evolutionary allegiance be clearly determined. As scale insects, they belong to the order called Hemiptera—true bugs, which are distinguished by their unique beak-like mouthparts. And the color? Squash one and a bright scarlet erupts from the delicate body.

Aside from the covering of fluffy wax, the famous red liquid is, ironically, the cochineal bug's main defense against predators. The liquid contains carminic acid, an anthraquinone that deters predatory ants. Two organisms not deterred are carnivorous caterpillars (*Laetilia coccidivora*) that specialize on scale insects and store the carminic acid for their own defensive purposes (they regurgitate the red liquid when attacked by predators), and humans. Once we brush the bugs off cactus pads, dry and pulverize them, we can treat the powder containing the acid different ways to produce different colors. If purified, the dye is often called carmine, but pseudonyms hiding cochineal's true identity include the somewhat innocuous "cochineal extract," the nebulous Natural Red 4, unsettlingly enigmatic Ci 75470, or the especially suspicious E 120. I will stick with the term cochineal. Cochineal from the species *Dactylopius coccus* almost entirely replaced inferior reds acquired from other natural sources. Cochineal was a superior combination of being easy to use, (relatively) plentiful, and, most important of all, of producing a stunningly deep crimson. As Edward Melillo reports in *The Butterfly Effect*:

Insects and the Making of the Modern World, cochineal has been used to bloody Macbeth's hands for the Shakespearean stage, and give British officers an extra-brilliant red coat. Coats of common soldiers were dyed with the inferior red of the lac insect, setting up a hierarchy further visualized with insects.

Today, cochineal is often selected as the go-to source of red, even when synthetic alternatives exist. Ever since Red Dye No. 2 was suspected of causing cancer in the 1970s and red edibles were pulled from shelves in the United States, fears associated with synthetic dyes have lingered. Instead of risking losing sales by using other synthetic reds, which are also scrutinized

Jennifer Angus's tiny, handmade books telling a story about cochineal bugs. Each is dyed with the pulverized bodies of cochineal bugs and comes with a glass vial of dried specimens, with the message "Open in Emergency" (1.7 × 2.4 in. / 4.3 × 6 cm).

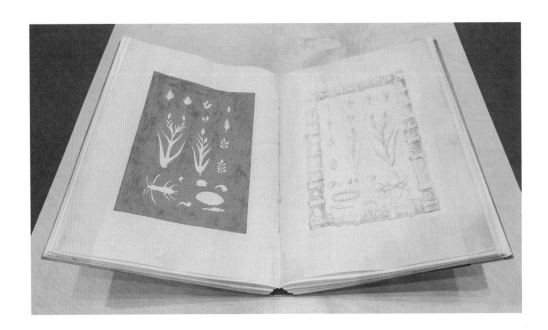

as posing possible health risks, some manufacturers of foods and cosmetics have made the switch back to natural sources of red. Cochineal has historically faced moderate competition in this arena. Some plants produce a deep red, like madder and sappanwood. The toxic mineral cinnabar is found in stoneware and lacquerware. Before cochineal from the New World spread to the East, red dyes were produced from other scale insects, including kermes (especially *Kermes vermilio*), Polish cochineal (*Porphyrophora polonica*), Armenian cochineal (*P. hamelii*), and lac insects (commonly *Kerria lacca*). Phoenicians traded kermes scale insects throughout the Eastern Hemisphere, and until Spanish ships laden with cochineal from the New World reached Europe's shores, red from other scale insects colored many textiles and illuminated manuscripts.

Spain's monopoly of the New World scale insect inevitably collapsed, and a shift in economics and influence followed. Somewhat symbolic of the monopoly

fading, the cochineal itself can fade. Exposed to light over time, the molecules containing cochineal pigment can disassemble and become translucent. Although some textiles and paintings remain rich with the reds of cochineal, others are not at all what they used to be, sometimes giving us a false impression of impressionists' works. Vincent van Gogh, fully aware that some of his paints faded over time, used cochineal lake (carminic acid from cochineal bugs precipitated on a metal salt) in all three versions of *The Bedroom* (1888–1889). Though he wrote "the walls are of a pale lilac" in a (translated) letter to his brother Theo, they now appear far bluer. Only recently have tests proven cases of cochineal use in paintings, and it wasn't only to produce reds, or violet mixtures for bedroom walls. Artists extracted cochineal dye from textiles to use as a glaze in their paintings, adding a shimmer and depth. Though evidence from faded paints and transparent glazes grows and informs us about historical cochineal use, it may be a long time before we grasp the extent to which museums' holdings are infused with the unstable dye of true bugs.

Permanence is difficult to come by. Art and artifacts fade and deteriorate with time. Impermanent pigments, paints, and inks are the bane of art curators' and conservators' lives. Finding a medium that will survive the ages means giving a document or a drawing the opportunity to send its unaltered message to future generations. One relatively stable ink owes its permanency to insects. Many gall wasps (for example, *Andricus kollari, Cynips quercusfolii*) lay their eggs in oak trees. Larvae hatch from these eggs and devour tiny bits of the oak from within, secreting chemicals that alter the plant's physiology as they go. The plant's defense is to build a protective growth around the damaged site, forming a gall. Plant galls are abnormal, tumor-like swellings that the invading parasite can use as shelter and food. Oak galls produce high concentrations of tannic acid, the same chemical that can make puddles brown when full of fallen oak leaves. Enterprising individuals experimented with these galls and their tannic acid to produce ink that lasts. Recipes vary, but this will give you a perfectly viable oak gall ink: (1) Pulverize the gall. (2) Boil it in water. (3) Add iron sulphate to create a blue-black color. (4) Add gum arabic (from an acacia tree) to make the ink smooth and help bind your ink to the writing surface. (5) Strain.

Oak gall ink was the Western world's go-to ink from medieval times until the late nineteenth century. Where can you find it? Start with Leonardo da Vinci's notebooks, Bach's musical scores, drawings by Rembrandt, Dürer, and Van Gogh, the Declaration of Independence, and the US Constitution. In the case of the Magna Carta, the document famously limiting the power of the English king, gall wasps represent one in a large community of organisms that played a role in its creation. The ink—caused by gall wasps' meddling with an oak, as described previously—was written on sheepskin with a quill made from a bird feather and sealed with silk cords using beeswax and plant resin (though only one of the four original surviving copies includes the seal). Note that along with two mammals (sheep and human scribe), one bird (probably a goose), and at least two plants (oak gall and resin source), at least three insect species (gall wasp, honey bee, domesticated silkworm) helped create and officiate one of the most important documents in history.

Magna Carta, the document that limited King John of England's powers and set a legal precedent that changed England forever, has a beeswax seal with silk cords. Only four originals of the 1215 document exist, and this burned copy, suffering from a botched restoration, is the only one with the Great Seal of King John still "intact," though King John's face is melted into an amorphous lump.

Now that we've dissected the biodiversity of the Magna Carta, it's time to consider the colorful questions posed earlier. Did you sit in the blue room? Seeing red does all manner of things to our physiology, including increasing our heart rate, but blue can lower our heart rate. The blue room may help calm you, depending on your previous experiences and associations with blue. Did you eat the camouflaged insect? If faced with something bright and gaudy in nature, it's probably best to err on the side of caution before consuming it. Many advertise their toxicity with bright colors or memorable patterns. Eating a flashy, colorful insect can leave you with a potent memory of what to forever avoid. Extreme caution, however, could mean passing up nourishing morsels in the form of palatable mimics of toxic insects. Finally, did you wear the red uniform? Wearing red has its advantages, but as with all of these examples that probe and prod our complex psychology, exceptions abound and context matters.

An Instrument's Hidden Insects

My sister, Korinthia Klein, is a luthier. She not only plays stringed instruments in orchestras and quartets, she also builds them. When my sister decided to become a musician and builder of violins and violas, she did not do so with insects in mind. Yet, the music she plays rises, flits, and falls with insect inspirations more often than she might have predicted. A magic swan-bird transforms a czar's son into a bee and he flies off to see his father in Nikolai Rimsky-Korsakov's "Flight of the Bumblebee." And there's Frédéric Chopin's Étude op. 25, no. 2 in F Minor ("The Bees"), and Béla Bartók's "From the Diary of a Fly."

When my sister assembles tools and building materials to construct a violin, the process becomes an exercise in applied botany and a study in biodiversity. Though synthetic instruments and "vegan" options exist, a violin, traditionally, represents the harmonic convergence of plants, mammals, insects, and the occasional mollusk or reptile. The bulk of a violin is made from woody plants, and features wood from maple (back, ribs, scroll, and bridge), spruce (top and soundpost), willow (internal blocks), ebony (saddle), ebony or rosewood (fingerboard), and pear (purfling), with pernambuco providing the density, strength, and sonic qualities for the stick of the bow. Glue made from rabbit skin or other mammal hide holds the violin together, and hair from a horse's tail is the

part of the bow used to glide across the instrument's strings. The strings are often made of metals or synthetics, but the traditional material used in the West is sheep or cow gut. Pulled while still hot from the carcass, the small intestine is dressed, stretched, twisted, dried, and polished. The bow's slide is often decorated with highly iridescent abalone shell. Though outlawed now, tortoise shell was sometimes used for the handle of the bow (inexplicably called the "frog"), and whale baleen was used for wrap to protect one end of the bow. Some bows still contain ivory, though crossing borders with such an instrument can land a luthier or musician in a Gordian knot of red tape. Sources of ivory include teeth and tusks, particularly of the proboscidean giants, whether they be vanishing or

A viola, made by luthier Korinthia Klein, is a symphonic synergy of organisms. Insects often factor in the creation of stringed instruments and their bows. A few products featured here: shellac, on neck (from lac insects on stick); cochineal, in varnish (from cochineal bugs on cactus); "bee sting," propolis, and beeswax (all from honey bees).

vanquished: the endangered African and Indian elephants or the extinct mammoths and mastodons.

This community of contributing animals would not be complete without arthropods. Earlier, I mentioned the novel use of silkworm moth silk and spider silk to make instruments, and spider silk to produce violin strings, but traditional materials of the luthier have long turned to insects. Varnish, that all-important veneer that holds such allure and mystery for violins built by Antonio Stradivari, history's most famous instrument maker, is often reddened by the pulverized bodies of cochineal bugs, those scale insects impregnating the red stripes of the first Star-Spangled Banner. Shellac, coating of choice for so many surfaces discussed previously, is used to protect the violin neck, and sometimes other parts of the body. Since shellac is reversible, it is perfect for repair and reworking an instrument. Propolis, collected from honey bee hives, is often used to coat the inside surfaces of a violin, just as bees coat the interior of their nests. Beeswax on the threading of the fine tuner screw can ease its turning and prevent rattling. Silkworm silk sometimes wraps one end of the bow instead of whale baleen. When purfling, that decorative inlay along the edges of a stringed instrument, comes together in a sharp point, this is called a "bee sting," so the very form of the instrument evokes insect anatomy. Insects also factor in when selecting bow hair and during instrument restoration. Mongolian and Siberian horses have to thwack fewer flies in their cold habitats, so their tail hair is valued for its thickness and condition when selecting bow hair. If your instrument languishes in its case, "woodworms" (wood-boring beetle larvae) can burrow through the wood and "bow bugs" (dermestid beetle larvae) will eat your bow hair. More incentive, as my sister astutely points out, to practice playing an instrument regularly.

As with the Magna Carta and the production of every other material object, insects perk up their antennae and tread with their tarsi in unexpected ways. The music that swells and stirs us from Korinthia Klein's instruments represents a sonic emergence channeled by the mingling of a multitude of organisms.

Paper

You may be flipping actual, tangible paper pages as you read this book. If so, these pages came from trees—trees that I have been assured were harvested sustainably and in a manner that did not appreciably exacerbate global climate change. Libraries, bookstores, and magazine aisles are stocked full of ex-trees, chopped, pulped and processed into reams of leaves. Trees are overwhelmingly the source for paper products the world over.

This wasn't always the case. The earliest writing surfaces included stone, parchment, and papyrus. By 150 BCE, people in China were macerating a mix of materials into the first paper, which spread westward. Recycled fibers from textiles were used in Europe. The rags were primarily linen, hemp, and cotton, but these supplies were running low. A need for another source of paper was pressing. Could we, yet again, turn to insects for inspiration?

Remember the French entomologist René Antoine Ferchault de Réaumur, who was tasked to investigate the viability of using spider silk to replace silk spun by silkworm moths? This same gentleman was invested in improving other industrial processes as well, and is credited with proposing a solution in 1719 to the paper shortage that was troubling Europe. Réaumur suggested we follow the ways of wasps and create paper as paper wasps do. "The rags from which we make our paper are not an economical material and every papermaker knows that this substance is becoming rare. While the consumption of paper increases every day, the production of linen remains about the same. In addition to this the foreign mills draw upon us for material. The wasp seems to teach us a means of overcoming these difficulties." Paper wasps gnaw wood and, layer by layer, build nests of "fine paper and as heavy as

Paper nest constructed by wasps with cutaway to reveal inner architecture, depicted by Edward Julius Detmold (*Fabre's Book of Insects*, 1921).

that of ordinary portfolios." Fast-forward more than a half century later, and Jacob Christian Schäffer adopted Réaumur's advice and experimented with a variety of materials for his 1765 book, the first of six volumes of paper trials. Among the trials, Schäffer included paper samples made from pine cones, green algae, and the nests of paper wasps.

Durable, the wasp nest paper was an indirect proof of concept that we could make paper from trees. After all, Schäffer had made paper *from paper* made from trees. If insects convert wood into paper, why couldn't we? Many species of wasps in temperate and tropical zones chew wood into a mulch, adding their saliva to produce exquisitely colored paper envelopes surrounding their perfectly hexagonal paper brood cells. These paper treasures hang from trees, are nestled within tree hollows, or are buried underground. Some form umbrella-shaped nests that hang

You are inside a dollhouse, constructed by Alastair and Fleur Mackie from sheets composed of about three hundred abandoned wasp nests, pulped and cut to match the coordinates of a wooden doll house kit.

from thin pedicels attached to eaves of buildings. The key to transforming wood into these architectural masterworks is to macerate with mandibles, add saliva, and build.

Wasps make it look easy, but creating paper from wood was expensive, and after failed attempts, machines were eventually built that could extract fibers and process paper using wood pulp. Although the Chinese had been using a variety of plant-based papers for almost two thousand years, it took another 124 years after Réaumur's suggestion to imitate the wasp before wood pulp became an economically viable, although less stable and environmentally problematic, source for paper.

Chitin

Traversing the cosmic void poses challenges. Let's imagine that circumstances have left you in a spacecraft, traveling with the aim of establishing a colony on a distant land. You have prepared for your planetary voyage by provisioning your craft with a recyclable source of oxygen and water, and sustainable sources of food. You brought bees to pollinate future agricultural plots, moths to produce silk, and flies and crickets to provide protein. You trained your mind and body to withstand the hazards of extended isolation and weightlessness. Despite all of your careful preparations, however, one threat looms. Residue from an exploding star created an enemy that is invisible and unrelenting. Galactic cosmic rays hurtle through space, as well as through your vulnerable vessel. All the layers of insulation won't stop the radiation now pummeling your body. How could you possibly protect yourself against such a cosmic menace? The irony—the salvation hidden in plain sight—may lie in the miniature cosmonauts joining you for your space voyage. The carcinogenic rays altering your body cells have little effect on your insect companions. Could you have coated your vessel with the exoskeletons of insects? Could you have fashioned a suit impervious to the damaging rays?

Insects and their armored relatives owe much of their success to the exoskeletons that protect them, and chitin is a key ingredient of this armor as well as in the cell walls of fungi. After cellulose, which fortifies the walls of plant cells, chitin is the most abundant biopolymer (a large molecule in nature composed of simple subunits linked together repetitively), and can be recovered in massive quantities from the crab and shrimp industry. Chitin can be rigid or flexible, is biodegradable, and has attractive, eco-friendly properties, and eco-friendly extraction methods are being developed, making it a material with manufacturing promise.

Chitin, when treated with an alkaline substance, can be converted to chitosan, and chitin or chitosan has potential applications in food, waste-water treatment, agriculture, human and veterinary medicine, cosmetics, and engineering. In food, chitin and its derivatives can emulsify, thicken, or stabilize components, combat microbes or gastritis, extend flavor, add dietary fiber, and more. For waste-water treatment, the positively charged chitosan attracts a variety of contaminants, and can recover heavy metal ions, pesticides, PCBs, and dyes from polluted waters. In the world of medicine, chitin in wound dressings and as vehicles for treatments can accelerate healing, and its flexibility and durability make it an attractive material for surgical thread and sutures.

One frontier of bioplastics involves molding, casting, and printing chitin from exoskeletons. Researchers at the Wyss Institute for Biologically Inspired Engineering have synthesized an alternative to petrochemical plastic called "shrilk" because it combines the attributes of chitin produced from discarded shrimp waste, and the fibroin protein of silk. It is inexpensive, biodegradable, lightweight, and compatible with living tissue. Unsurprisingly, shrilk succeeds because it approximates not only the composition of an insect's cuticle (the hard, nonliving part of the exoskeleton), but also its structure. Like layers of plywood, insect cuticle's strength is bolstered by a laminar design.

Shrimp, crabs, and other marine crustaceans have been a source of chitin for this type of research and development, but soldier flies, mealworms, rhinoceros beetles, grasshoppers, stinkbugs, cockroaches, and silkworm moth pupae are but a few of the insects from which chitin has been extracted and scientifically tested with industrial applications in mind. Even chitin-packed tarantula moltings have been the subject of inquiry and study. When spiders grow, they shed their exoskeletons, and from this rich source of chitin, alone, people have created scaffolds for tissue engineering. Replacing damaged tissue, or creating new tissue in our bodies, could benefit from the molted exoskeletons of others.

Chitin is a versatile material and, if by some fluke, you and your insect companions make it to your planetary destination, chitin may be a critical component to your survival upon arrival. You have grand plans to scale up your colonies of insects to serve as pollinators, soil aerators, decomposers, and as food. You will harvest their silk, honey, and wax. But building facilities to house such operations in a resource-starved environment poses a daunting challenge. Ng Shiwei and colleagues in Singapore have devised a way to mix chitin from exoskeletons with extraterrestrial soil to form a building material that could help support humans on Mars, or beyond. Interplanetary colonization seems inevitable, but to make extended stays on other lands possible, Shiwei and colleagues saw a need for critical technologies to mature quickly, and created a building material that relies on accessible and sustainable materials. The concoction consists of chitin and a surrogate for regolith—a combination of dust, broken rocks, and other material covering terrestrial planets and moons. Treated with sodium hydroxide, acetic acid, and water, the resulting "biolith" is comparable in mechanical properties to refractory brick (ceramic used to line the floors of furnaces, kilns, and chimneys). As a proof of concept, the researchers used a mold to cast a functional wrench made of the biolith, and applied the biolith as a mortar to hold together a model of a Martian habitat module. Since insects will likely be a key sustainable source of food in any artificial environment we establish away from Earth, and the chitin that makes up much of their bodies isn't a source of nutrition for us, we could extract chitin without diminishing a primary source of our food. Insects and their armor could be the key to our interplanetary and planetary survival.

Insects in Space

Though chitin-coated spacecraft is science fiction today, insects (and arachnids) are regulars when it comes to space travel. The first animals to ever enter space weren't dogs or monkeys, but fruit flies. Fruit flies and plant seeds boarded German V-2 rockets, claimed by the United States following Germany's fall after World War II, and launched from New Mexico's White Sands Missile Range in 1946. Fruit flies again pioneered space travel via a V-2 rocket launched on 20 February 1947, this time by returning to Earth alive, demonstrating for the first time that animals can survive a journey to space. These dipteran astronauts flew for 190 seconds at a height of 68 miles (109 km; 62 miles [100 km] above sea level is used by space treaties as the beginning of outer space). Moss, mice, monkeys, dogs, and others followed, until Yuri Gagarin became the first human to enter space, fourteen years after the flight of the fruit flies. Other fruit flies, and parasitic wasps, flour beetles, crickets, and spongy moth eggs, were among other test subjects sent into space in the 1960s. The very first earthlings to enter deep space or to orbit the moon included mealworms (beetle larvae) and more flies, and were joined by two Horsfield's tortoises aboard the Soviet *Zond 5* in 1968. Another heroic first came in the form of a cockroach named Nadezhda, who became the first earthling to give birth to organisms conceived in space (aboard the European Space Agency's *FOTON-M3* in 2007).

Mars Beetle from the *Celestial Cabinet* by Karen Anne Klein

Initially, researchers flung insects, spiders, and others into space simply to see if they would make it back alive and unharmed. Radiation exposure at high elevations is a danger, and testing nonhumans during the space race was an imperative. Many insects and spiders survived, but some did not. The tragic, final flight of the Space Shuttle *Challenger* in 1986 carried silkworms, garden orb spiders, carpenter bees, and harvester ants, among other organisms. Only nematode roundworms were recovered alive among the debris of the fallen shuttle. Later missions have since tested, and continue to test, for behavior in zero gravity, and as preparation for use of other species to realize our long-range ambitions in space.

Venom and Poison

There are gods, and they do throw thunderbolts.
Poseidon just rammed his trident into your breast.

Disappointing. A paper clip falls on your bare foot.

As poetically conveyed in these quotes, Justin Schmidt reported the sensations he felt when stung by eighty-three different species of wasps, ants, and bees. Not all stings are created equal, and when Schmidt ranked the pain he felt when envenomated by his perturbed subjects, the Schmidt sting pain index was born. The paperclip is the sting of a club-horned wasp (*Sapyga pumila*). Only a 0.5, this pales in comparison to Poseidon's trident, the sting of a giant paper wasp (*Megapolistes* sp.), ranked by Schmidt as a 3. Even a god's trident apparently doesn't approximate the effects of the tarantula hawk wasp (*Pepsis* spp.), warrior wasp (*Synoeca septentrionalis*), or bullet ant (*Paraponera clavata*), each of which earned a top ranking of 4.

The sting of an insect can make for a memorable experience, especially if it is accompanied by excruciating pain. From an insect's perspective, causing pain can be a life-saving defense strategy. Some insects invest a lot into building a towering nest, provisioning their nest with food, or protecting their young. When you live alongside enough hungry invaders, you evolve defenses, or fade into oblivion. If somebody came along and ripped open your house, menacingly lunged toward members of your family, or began to crush your body, injecting venom could be a handy tactic for persuading the perpetrator to reconsider their threatening ways. Delivering the right concoction of venom to an appropriate menace could save your life . . . or preserve your genetic legacy.

Venomous animals tend to inspire fear, and there is reason why fear might be a valuable default response. Some venoms are not only painful but they can also be deadly. If your first impulse is to do anything in your power to avoid venoms, few would fault you. But what if you could harness these natural concoctions, which harbor such clear efficacy and potency, to your advantage? Human applications for venoms exist and, in the right hands, venoms can be used to kill, cure, or transform.

To Kill

A venom is a toxin actively injected by an animal. Think of the pain-inducing liquids injected by a wasp's stinger, a snake's fangs, or the many nematocysts of a jellyfish's tentacle. A poison, on the other hand, is a toxin passively transmitted, so you would have to go out of your way to suffer nature's chemical wrath from a poisonous animal. The San people of Southern Africa have devised a way to use poisons when hunting for food, and insects and arachnids can be key to their survival. Drawing from their broad and deep knowledge of ecology, the San sample a wealth of pharmaceuticals from specific plants and animals, and prepare for a hunting expedition by lacing their weapons with venoms from scorpions, spiders, or snakes as well as poisons from beetles. The beetles, or their worm-like larvae, are found on poisonous plants, as first described by the Finnish explorer Hendrik Jacob Wikar in his 1779 journal:

> The two kinds of tree poison they use on their arrows are the most
> remarkable of all. The tree producing the first, the strongest poison, grows
> in the mountains along the Great River and has a very powerful scent, so
> different from that of all other trees that one can find it by its scent alone
> without knowing anything more about it; its foliage is green. In July the
> poison worms, which during the time that the tree is dry live at the bottom
> in the grey-brown bark of the stem, begin to appear on the leaves. These
> worms are exactly the same colour as the leaves they eat. [People] take
> only the worms which are tightly tied up in a piece of leather and kept
> until they rot. Then they are ground to a fine powder which is rubbed all

Preparing poison arrows with beetles, by the Juǀ'hoansi in Tswumke Conservancy, Namibia (Kalahari region). Trechie is collecting beetle cocoons from the deep sand at the base of a *Commiphora* shrub.

ABOVE RIGHT: Cocoons of *Diamphidia nigroornata* beetles

round the arrows with spit. No one wounded with this mixture, when the gall of the big rock lizard has been added, has any chance of recovery, unless he is immediately given the urine of a poison drinker. When a twig of this tree is broken off a strong-smelling sap or oil oozes out. You must be careful that this does not get into your eyes, for if it should you would become stone blind.

Wikar follows this with a harrowing tale about how he ate honey produced by bees collecting nectar from the same plant. A local companion had offered the honey for him to create beer, and warned Wikar not to eat it. Wikar paid no heed, and after tasting a spoonful, his throat burned like fire, he vomited violently, and fell unconscious. A fortuitous side effect: He had suffered from tapeworms since childhood, and the poison-laced honey left his body so uninviting a home, that even these internal parasites, "three fathom long" (18 feet / 5.5 m), or longer, had to evacuate the premises.

The distinction between venom and poison becomes fuzzy when poisons found in nature are actively delivered into prey via makeshift stingers, in the form of the San's poisoned spears or arrowheads. To poison their weapons, the San dig up to about ten feet (three m) underground to retrieve the larvae of different species of leaf beetles or ground beetles, squeeze the larvae (or

let them dry, then pulverize them), and mix and mash their innards with human saliva or plant extracts into a paste. They rub this cocktail onto the shaft behind the pointed end of their weapon. How potent is the mix? It is slow-acting, so large animals can take three days to succumb, but it is effective. Even their largest prey slow, stumble, and fall victim to the toxic mélange. The alkaloids from plants can last ages—arrows acquired in 1806 were unchanged in their toxicity when tested eighty-eight years later. Beetle poison has a shorter shelf life, but with a report of potency after two years, it pays to be cautious when handling old weapons laced with the beetles' innards. Presumably having such a toxic makeup helps defend the beetles against predators, but no one has yet figured out for certain how these beetles might benefit from having such poison in their bodies.

ABOVE LEFT: Squeezing contents of beetle larva onto giraffe bone to prepare arrow poison

San quiver and poison-coated arrows. Each arrow has a poison-covered tip made of bone; the darker arrow tip (*back*) is completely covered with poison. The quiver is wood with skincap at each end, bound by hide strips and with leather carrying strap.

To Cure

We can use toxins to kill, but as Wikar's fortuitous tapeworm expulsion so gloriously demonstrates, toxins can also cure. An example that has survived the ages is well known to beekeepers. For thousands of years, people have prescribed honey bee venom as a remedy for a wide variety of ailments. Bee venom therapy found its origins in ancient China, Egypt, and Greece, with

Hippocrates among those using honey bee venom in their medical practice. Today, honey bee venom is delivered by way of injections, acupuncture, ointments, cream, ultrasound gel, or the all-natural method of coercing bees to sting the afflicted site. Though bee venom can have side effects, and some risk experiencing anaphylaxis, bee venom has beneficial anti-inflammatory and, ironically, pain-relieving properties that have attracted many a medical investigation. Side effects aside, studies involving randomized controlled trials using honey bee venom cite improvements for arthritis, arthralgia, Parkinson's disease, lower back pain, temporomandibular disorder, and delayed-onset muscle soreness. Other studies suggest bee venom has antimicrobial, anti-oxidant, antimutagenic, and anticancer properties, with potential to combat Alzheimer's disease, multiple sclerosis, atherosclerosis, and amyotrophic lateral sclerosis (ALS).

If so much potential exists in the venom of one species of stinging insect, it pays to shop around. Each venom is a different cocktail of chemicals—sometimes thousands of chemicals—so the prospect of finding a venom that is better suited to combatting one neurodegenerative disease, cancer, or other malady than another lends biomedical value to exploring biodiversity. Sure enough, promising tests exist with venoms of ants and wasps (especially hornets). Poisonous proteins from moth caterpillars show cancer-fighting activity, as do the poisons found in cantharid beetles. On the arachnid front, some spider and scorpion venoms also show anticancer potential. Venom from the Australian funnel-web spider can help with cardiovascular disease by reducing cell death after a heart attack. Venom from the Israeli death-stalker scorpion can uncover and illuminate breast, brain, and colon tumors.

A first step in prospecting for chemicals that holds medical potential is to look for organisms with chemical defenses. Another approach is to consult shamans and others with knowledge of traditional practices for insights about natural sources of medicinal relevance. If chemicals can be tested and synthesized, a path to building a pharmacy with insect or arachnid origins could transform how humans heal.

Ours is not the only culture, or medicine cabinet, affected by insects. A chimpanzee troop in Gabon appears to be using some mystery insect

for possible self-medication. Members of the troop have been spotted grabbing small, winged insects from the undersides of leaves and rubbing them in their open wounds, holding an insect between their lips and repeatedly transferring the insect to their injury, or the injury of a fellow member of the troop. What function does this serve? Does the insect produce chemicals that relieve pain or dismantle harmful microbes? Can the chimpanzee distinguish this insect from closely related species? Are there many insects in the chimpanzees' surroundings that contribute to their health? How many missed opportunities are there for chimpanzees, or us, to use insects for medicinal purposes? Documenting insect medical practices in nature may reveal a surprising breadth of the innovation in other animals, with untapped applications for us.

Thea, an adult male chimpanzee in Gabon, is examining an insect, which he is about to use for possible self-medication purposes.

To Transform

Toxins can kill or cure, but there are other ways in which toxins can transform us. Insect toxins can initiate you into adulthood, take you on a spiritual or recreational journey, or stimulate your sex drive. Let's begin with an insect venom–fueled rite of passage.

Trial by Venom

From thunderbolts to paperclips, we initiated our toxic journey with a pain index of insect stings. Three stood out as exceptionally excruciating, and the one most famously feared is that delivered by the bullet ant, *Paraponera clavata*. Solitary, robust hunters in Central America and the Amazon basin, these ants squeak when grabbed, can pinch with their mandibles, but it is their sting that has placed them in a special category of fearful reverence. Due to their infamy, this ant is showered with many common names, including "bala" (bullet, because the pain of the sting, for some, is akin to being shot) or "hormiga ventiquattro," referring to the twenty-four hours it can take for the pain to go away. The bala is a common focus of attention in the geographic region where they live, and attaining manhood, for some, can hinge on a ritual where a boy becomes a pincushion for bala stingers.

The Sateré-Mawé people live in the Brazilian Amazon and continue an insect tradition that distinguishes them from all other communities in the rainforest. When a Sateré-Mawé boy wishes to transition into manhood, he can volunteer to undergo a ritual (Waumat) that demonstrates his tenacity in the face of extreme suffering. First, a bullet ant colony is found at the base of a tree, and ant workers defending their nest are gathered and anesthetized in an herbal brew. The temporarily calmed ants are woven into mitts made from palm fronds so that the stingers face inward. These mitts are then inserted into ceremonial gloves also made of palm. Once the ants revive, the boy faces the agony and temporary paralysis of dozens of ants injecting their venom, and, if sufficiently stoic for five to ten minutes (or for thirty minutes according to one report), the boy is one step closer to being eligible for marriage or leadership roles in his community. Additional, mandatory steps: Repeat the ritual for a total of twenty to twenty-five such tests over the course of months or years. Each stinging session can leave the envenomated boy shaking uncontrollably for days.

A related coming-of-age stinging ordeal that has recently faded from practice is the Wayana (Oyana, Ajana, etc.), ëputop, or maraké, and uses either bala ants, or *Polybia* wasps. Boys are fitted with a breast plate woven

Stinging insects were woven into plaques of palm and worn by Wayana boys during a coming-of-age ceremony. The Wayana used bala ants (*left images*, both from Pará, Upper Paru de Leste River, Brazil) or wasps (*right images*; *Polybia liliacea*, from French Guiana), with the business end of the stinging insects (*bottom images*) facing the boys' bodies.

with ants or wasps, and wear these stinger-packed breast plates for what must be a very long night.

Biologists studying in tropical America can sometimes be drawn to the sting of the bala in what resonates as an echo of the rite of passage exhibited by the Sateré-Mawé. Weathering a sting can serve as a badge of courage for tropical biologists who pride themselves on surviving grueling bouts of parasitism or other natural tribulations—like hosting bot flies. Some bot flies grab other insects, lay eggs on them, and when these insects come into contact with a large, warm mammal, the bot fly larvae hatch and burrow into that mammal's flesh. Sometimes that large, warm mammal is a human. As the fly larvae develop and grow in size, their spines grow in kind, so a turning movement by a feeding, spiny bot fly larva can make life uncomfortable for their human host. If you have the stamina, you can carry larvae through their third "trimester" until they naturally emerge to squirm away and continue their metamorphosis elsewhere.

Why the deviant urge? Bot flies and bullet ants may give some of us a symbolic test. A drive to overcome adversity may be deeply ingrained in us. The arc of every story relies on conflict, and overcoming adversity is the mark of any hero. For some, insects present a hardship to endure or a cultural test for proof of our worth, or our station in society. To know joy, it may be important to experience its antithesis.

Mind Trip

Our brains are awash in chemicals, and it isn't difficult to alter the balance of these chemicals, or to introduce entirely new chemicals into the mix. Some of these chemicals modulate how our nervous system operates and can change how we perceive the world. Was that insect buzzing in your ear or scurrying across the floor real or a product of your mind? Do you feel invisible insects crawling on your body? "Formica" is Latin for ant: Formicanopia is the visual hallucination of insects and formication is the imaginary sensation that insects are on or under the skin. These tricks of the mind can become serious, and if you are convinced that the imaginary insects on you are real, delusional parasitosis has set in.

Just as mysterious as fictive insects causing mental and physical suffering, our minds can respond to pain delivered by real insects in strange and individual ways. Justin Schmidt describes the sting of the bullet ant "Like walking over flaming charcoal with a 3-inch nail embedded in your heel." Though such a sting is expected to bring agony to any human recipient, a person's perception of the pain depends on factors like placement of the sting or amount of venom injected, and varies person to person. Your mind might cope with the pain by embracing the experience, or teleporting to a realm of distractions—where anything *but* ants captures your attention. Alternatively, you might transform the ants into creatures you could battle, or something innocuous, something to tame.

One way to escape reality or expand one's consciousness is by harnessing psychotropics from nature. Fungi (magic mushrooms), plants (ayahuasca), fish, and frogs are the primary sources for such drugs, but one bird (oconenetl, eaten by the Aztecs) and a few insects have escorted humans through history on trips of the incorporeal variety. Though psychotropic experiences with insects appear to have vanished, their existence is worth mentioning so we can have a better appreciation of lost Indigenous culture, and in light of recent findings that show medical benefits of mind-altering drugs. The first published account about a hallucinogenic insect comes from Augustin de Saint-Hilaire (1824). The caterpillar of a moth, possibly *Myelobia* (*Morpheis*) *smerintha*, was used by the Malalis in Brazil as a source of food and medicine . . . and occasionally something more:

> When I was among the Malalis, in the province of Mines, they spoke much of a grub which they regarded as a delicious food, and which is called bicho de tacuara (bamboo-worm), because it is found in the stems of bamboos, but only when these bear flowers.

> Some Portugese [sic] . . . value these worms no less than the natives themselves; they melt them on the fire, forming them into an oily mass, and so preserve them for use in the preparation of food. The Malalis consider the head of the bicho de tacuara as a dangerous poison; but all agree in saying

that this creature, dried and reduced to powder constitutes a powerful vulnerary (for the healing of wounds) . . . it is not only for this use that the former preserve the bicho de tacuara. When strong emotion makes them sleepless, they swallow, they say, one of these worms dried, without the head but with the intestinal tube; and then they fall into a kind of ecstatic sleep, which often lasts more than a day . . . They tell, on awakening, of marvellous dreams; they saw splendid forests, they ate delicious fruits, they killed without difficulty the most choice game; but these Malalis add that they take care to indulge only rarely in this debilitating kind of pleasure.

This centuries-old, unscientific note is likely one among a number of scattered, disappearing anecdotal accounts of insect hallucinogens. A more carefully documented case comes from a painful ant encounter quite different from the ritual of the Sateré-Mawé. Seventeen Indigenous cultures across south-central California traditionally used harvester ants (*Pogonomyrmex californicus*) medicinally. Seven of these cultures, most notably a handful of Shoshonean groups, used the ants to acquire supernatural, shamanic powers in the form of "dream helpers." The ants themselves were viewed as manifestations of supernatural power.

To become a dream helper meant consuming massive quantities of live harvester ants, after several days of fasting and vomiting, with a postmenopausal woman to guide the ritual. What was her job? Take the subject away from the village, lay him on his back, and find a nest mound of ants. She let four or five ants cling to a ball of moistened eagle down, then inserted this ball into the mouth of her subject, who sucked in ball after ball after ball of ants and feathers until he could take no more. After he consumed some eye-popping number (an account recorded by José Juan Olivas mentioned fifty or ninety balls as possible quantities), she startled him so that the ants stung and bit his innards in synchrony. The impact of these stings from within his body rendered him suddenly unconscious. Considered a small death, this catatonic state was accompanied by visionary experiences.

Depending on the stamina of the subject, the ritual could repeat for days, without food, though repeated administrations were not typical. Once conscious, vomiting was induced by drinking hot water, and harvester ants were said to be regurgitated alive. What was the purported payoff? If successful, the ant-envenomated subject attained powers, which could enable him to evade arrows, cure others, make rain, or transform into a different animal, for starters. Whether the venom alone was responsible for causing hallucinations is unknown. Peter Groark, who researched this practice, was careful to note that the hallucinations may have arisen as a side effect of physical stress from fasting and being exposed to the elements, as well as from a tribe member's cultural preconditioning. These factors, coupled with the shock of absorbing copious quantities of venom, likely augmented the visionary venture.

Another mind-altering experience guided by insects begins with what, at first, appears to be a sweet and innocuous treat—no balls of ants and waves of envenomation. Honey, that most beloved of sweeteners, can come with surprises, depending on the flowers from which honey bees collect their nectar. If honey bees happen to collect nectar from rhododendrons, which contain neurotoxins (grayanotoxins), they will produce the infamous "mad honey," which induces hallucinations. People eat mad honey for therapeutic reasons, recreational reasons, and, despite the risk of suffering from food poisoning or worse, to increase their sexual arousal.

Sexual Arousal

Love potions are often a snake charmer's ruse that too frequently place animals at risk of exploitation or extinction. Pulverized rhinoceros horn or a bowl of tiger penis soup are not the key to stimulating your sexual appetite, yet the mammals slaughtered for these mythically medicinal or erroneously aphrodisiacal parts are quickly vanishing. The hunt for aphrodisiacs knows no limits, and insects, though too rarely the focus of our appreciation, produce chemicals used to stimulate our desire. Victor Benno Meyer-Rochow, prolific and itinerant biologist, compiled a comparative survey of practices using animals without backbones for medicinal reasons, and stumbled upon examples of arthropods used as aphrodisiacs. Cure impotence? Try a darkling beetle called the "love bug" in Brazil, a male silkworm moth in South Korea, or ants in Colombia and Brazil. Boost sexual stamina or prowess? Ants in China or termite kings in sub-Saharan Africa. Aphrodisiacs? The ash of silkworm moth cocoons and larvae, or members of multiple beetle families. An "ant Viagra" is made of fresh weaver ants (*Oecophylla longinoda*), mixed with the ground seeds of two tree species, and sold in bars and dance clubs in Cameroon. Arachnid aphrodisiacs include a red velvet mite, or the ashes of a sheet spider's web (plus honey), both from India. According to a collection of ancient and modern Chinese medicinals, a sexual tonic stimulating erections can be derived from a stinkbug (*Aspongopus chinensis*), an aphrodisiac from a cicada (*Huechys sanguinea*), and sexual stimulants from dragonflies. Of the same dragonflies, however, the *Po-wu Chih* by Chang Hua also says, "that if the heads of dragon flies be taken on the fifth day of the fifth moon and buried inside the house they can change into azure pearls." As if in anticipation of possible skepticism regarding this claim, the following sentence reads, "It is not known whether this is so or not."

The most famous insect aphrodisiac is a poison secreted by a beetle with the misleading name "Spanish fly." This beetle exudes danger throughout its development. The beetle's eggs are covered in blister-producing cantharidin—classified as an extremely hazardous substance in the United States. Larvae hatch from the eggs, climb flowering plants, and latch onto unsuspecting solitary bees visiting the flowers, to be carried to their nests where the larvae devour the bee's offspring, as well as their food. After the beetles emerge as adults from the solitary nests, they glisten a metallic and iridescent green-gold, exquisitely advertising their toxicity. Powdered cantharidin

can cause physical effects that appear to mimic sexual arousal, but a delicate balance between arousal and death has loomed over its use since antiquity. The Spanish fly's warning coloration, ironically, attracted many people wishing to exploit their poisons, resulting in centuries of erotic forays, painful mishandlings, and fatal

Lytta vesicatoria is the "Spanish fly" historically used as an aphrodisiac for risky erotic forays. (southern France)

mishaps. Its very name, *Lytta vesicatoria*, means "raging madness + blister," which should tip off the astute polyglot. Then again, we tend to throw caution to the wind, if the rituals, victuals, and medicinals presented earlier are representative of our relationship with toxins.

USING INSECT BODIES

Insects secrete and spew, color and coat, emit and exude a cornucopia of chemicals, though we've only touched on a tiny subset. Those same bodies responsible for creating the silks, waxes, coloring agents, lacquers, and toxins have successfully shuffled and tinkered with the periodic table of elements long enough to serve as promising candidates for medicines, and for cuisine. We have used their bodies to medically treat and sustainably eat as long as we have existed as a species.

When not crushed or consumed, insects are durable, and a resourceful person can use their wee bodies in big ways. Their very presence offers useful clues to the receptive and perceptive investigator. If you wish to gauge the health of streams, look for caddisflies, mayflies, or stoneflies. If your aim is to determine when a person shed their mortal coil, identify the maggots or grubs on the carcass. If insects are present, but displaying unusual physical anomalies, you can use them as indicators of biohazards.

The former Soviet Union's Chernobyl Nuclear Power Plant meltdown in 1986 caused gross deformities in a variety of animals, with radiation-battered impacts documented in fruit flies, stag beetles, leaf beetles, and true bugs. Butterflies have similarly suffered severe abnormalities following the Fukushima Daiichi Nuclear Power Plant catastrophe in 2011. Even functional nuclear power plants appear connected to insect aberrations, according to work conducted by artist and activist Cornelia Hesse-Honegger. Mouthparts and antennae are askew, wings are deformed, and other asymmetries serve as reminders that human-modified habitats can harbor horrors.

An alternative to the observational approach (looking for insects in the environment) is to put insects to work, training them to monitor the safety of

Cornelia Hesse-Honegger, Garden bug (*Raphigaster nebulosa*)

environments. Like Pavlov's dogs, drooling to the sound of a bell (that at one time accompanied food), you can train moths, grasshoppers, parasitic wasps, or honey bees to respond to the smell of just about anything in the world. Ants can be trained to detect cancer cells. Culture cancer cells in a medium, train ants to associate the

An asymmetrical shield bug (*Rhaphigaster nebulosa*), found near the Leibstadt Nuclear Power Plant, Germany, and painted by Cornelia Hesse-Honegger

smell with a sugar water reward, and when presented with the same medium with cancer cells, without cancer cells, and with a novel odor, trained ants will spend more time in the vicinity of the previously rewarded cancer cell medium. Honey bees are particularly promising because hundreds of studies have been published since the inception, in 1961, of a standardized method of training bees to extend their tongue-like proboscis when trained to a stimulus. Harness a single bee in a small tube, touch a drop of sugar water to her antenna, and she will extend her proboscis to imbibe food. Feed her the sugar water while simultaneously puffing an odor across her antennae. Repeat. When she responds by extending her proboscis with the puff of odor alone, she has been trained to that odor. The beauty of training honey bees is that they can be trained to multiple odors, put into action in one to two days, cost less than sniffer-dog teams, are too light to trigger mines, and beekeepers are everywhere. Tests show promise for employing honey bees to detect landmines, bombs, explosives, as well as tuberculosis and COVID-19.

Honey bees are covered in branched hairs that become laced with traces of the environment, so they can serve as biosensors to track regional contaminants. Powder the hive entrance with a fungus, and the bees act as crop dusters, delivering the fungus that combats plant pathogens. String beehives along a fence and you can ward away African elephants from trampling your crops. Feed black soldier flies (*Hermetia illucens*) organic wastes and process biofuels superior to rapeseed oil–based biodiesel. Even our pernicious plastic bags and Styrofoam can be eaten by waxworms (*Galleria mellonella*) and mealworms (*Tenebrio molitor*), with some evidence that the plastics are chemically degraded in the process.

And then there are the ecological services that insects so vitally perform. We transport honey bee hives to pollinate crops, sometimes driving them thousands of miles to set up shop where they are needed. We attract and house other insects to promote pollination in our gardens and fields. We introduce exotic species to decompose dung, or to parasitize other exotic organisms that have run amok because no native predators or parasites have evolved alongside them.

In addition to ecological services, we erect butterfly gardens to appreciate insects' beauty, and benefit from ecotourism associated with the celebration of migrating monarchs. We educate with insects. We conduct science with insects, learning how nature operates, and about ourselves, due in no small part to our shared DNA inherited from an ancient common ancestor. The vast collections of insects held in museums ceaselessly guide and inform scientists wishing to learn more about the climate, evolution, ecology, and behavior.

Sword-hilt collar and pommel (*fuchi-gashira*) with cicada nymph and adult in copper-gilt alloy and gold; late eighteenth century Japanese

On a far darker note, we have exploited insects in gruesome ways during times of war or for purposes of conducting torture and spreading misery. With detailed accounts of Japan's development of bombs harboring millions of plague-infested fleas during World War II to evidence of US biological warfare using insects as vectors, Jeffrey Lockwood does this horrifying topic grim justice in his book *Six-Legged Soldiers: Using Insects as Weapons of War*. Because select insects can vector disease so effectively, like lice with bacteria causing epidemic typhus, fleas with bacteria causing bubonic plague, mosquitoes with viruses causing yellow fever or dengue, and mosquitoes with single-celled *Plasmodium* causing malaria, weaponizing insects became an attractive option to crafters of war campaigns. Even today, fears of terrorists using insects to spread disease or harm agriculture compel federal agencies to devise counterterrorist plans specific to these possible threats.

What follows are examples of how we use insect bodies in our cultural practices. Our diets, medical toolkits, and art are bolstered by a barrage of beetles, cadre of caterpillars, and army of ants.

Food

Hemiptera crews—
spittlebug kin—
with see-through wings, sing
their visceral joys & rues,
their static-brittle noise, their
emphatic dog days' din.

—BILL HARRIS ("CICADAS"
FROM *TINY BEASTS*, 2013)

A Princeton cemetery is teeming with life. Among the remains of humans, ritually buried with their weathered names and dates chiseled in stone markers aboveground, the earth is rustling, shifting, churning, and bursting with the bodies of millions of periodical cicadas. Entombed, but very much alive, cicadas suck fluids from plant roots as they slowly mature (cicadas belong to the order of "true bugs," fashioned with a piercing, sucking proboscis).

Of the 3,324 described species of cicadas at the time of this writing, nine of these are periodical cicadas, meaning that they synchronously emerge as adults after prolonged periods of developing underground. The World Cup cicada (*Chremistica ribhoi*), long known to villagers in northeast India, but only scientifically described in 2013, emerges every four years, coinciding with the same years the world goes wild over football/soccer. A second species, *Raiateana knowlesi*, emerges every eight years in Fiji. The remaining seven species all live in the eastern United States and remain underground for either thirteen or seventeen years (*Magicicada* spp.).

This monotonous subterranean existence amounts to as much as 95.5 percent of their unusually long lives, spent in quiet, absolute darkness. As long as no one has paved the earth above them while they develop below, the cicadas emerge to molt, spread wings, and live out their remaining four to six weeks in a reproductive frenzy. Too many surface simultaneously in preparation for the frenzy for predators to consume them all.

But why the thirteen- or seventeen-year wait? The reason for their predictable, protracted periods underground is a mystery. The origin may relate to surviving extended cold spells or variable weather conditions during the Pleistocene, about one to two million years ago. Why these extended burials continue to exist has probably been debated since at least 1715, when Reverend Andreas Sandel wrote in his journal about a mass emergence of cicadas. Because emerging together in great numbers overwhelms predators and parasites, making it impossible to exterminate an entire brood of cicadas, the function of the long development time might allow slower cicadas to catch up so all can emerge more or less synchronously. Also, if you spend a large, prime number of years hidden away, it might be difficult for predators or parasites to synchronize their life cycles with yours, and to target you as a host. A species that depends on periodical cicadas as a food source, but cannot sync up with their life cycle, will not be long for this world.

Only one species of fungus, *Massospora cicadina*, is known to coincide with the periodical cicada's life cycle. It has cracked the periodical cicada's code, with resting spores infecting male and female nymphs as they burrow and first emerge from the earth. The fungus spreads, infiltrating their adult bodies, the ends of the now-infertile cicadas' abdomens fall off, and a chalky plug disperses new spores that will infect the next generation of nymphs. Males sing to females, even when infected, but their songs have a higher pitch as their bodies are filling with fungi, so they sound smaller than they actually are to the ears of listening females. As devastatingly effective as this fungus is at infecting its host and altering cicada behavior, even this specialist cannot wipe out an entire brood of cicadas.

A mass emergence of cicadas is breathtaking and ear-splitting. It is a wonder of nature that marks time with auditory flourish and drama. If you

cannot hold out for the passing of Halley's Comet (as Mark Twain did), you can probably at least look forward to one or more periodical cicada emergences. There are many ways of celebrating these events. Gene Kritsky scientifically monitors cicadas' life cycles and promotes crowd-sourcing their sightings through the application Cicada Safari. Jennifer Angus creates art installations on massive scales that feature cicadas on walls, in cabinets, and under glass domes. David Rothenberg composes music and collaboratively plays outdoors with the screeching hordes. Joseph Yoon prepares them for supper.

Joseph Yoon is picking cicadas (not infested with fungi) off gravestones. He is a chef on a ceaseless quest to expand flavor profiles by including a diversity of insects in human diets. Although cicadas have served as symbols of resurrection, of longevity, and of happiness, and are best known for their courtship choruses, they also add depth to a tasty kimchi. Cicadas have been traditionally eaten by humans in different parts of the world, including where all nine of the periodical cicada species breed, and the cemetery is a profitable site for harvesting Brood X—one of the seventeen-year cicada populations.

Emerging after eight years in the earth, Fiji's periodical cicada (nanai, *Raiateana knowlesi*) was celebrated on the one hundred banknote from 2012–2020, paired with smiling faces, cruise ship, and snorkeler (*reverse side*). Insects have adorned banknotes issued by Algeria, Morocco, Uganda, Zimbabwe, Comoros, Papua New Guinea, Solomon Islands, Cook Islands, Seychelles, Malaysia, Bhutan, Thailand, Kazakhstan, Kyrgyzstan, Switzerland, Czech Republic, Canada, Mexico, Costa Rica, Suriname, Australia, and other entomophilic countries.

One lone periodical cicada (*Magicicada cas-sini*), experiencing adulthood after seventeen years in the earth. A member of Brood XIV, one of fifteen broods of cicadas that occur in the eastern United States, this cicada emerged with plans to mate in 2008.

Yoon has been preparing for this moment for months, choosing the time and place for collecting cicadas after consulting experts and historical records. Planning can be important when human construction has limited the historical range of the chorus and natural spectacle of periodical cicadas. Attempting to time his harvest, Yoon had to follow the weather closely, then forgo a normal sleep schedule to take advantage of the ephemeral, explosive emergence. He freezes tens of thousands of cicadas in preparation for future ventures as an entomophagy ambassador and spokesperson for food safety and sustainability. As human populations grow unsustainably, more careful thought must be paid to secure food throughout the world, and insects are often seen as a means of helping attain this security. Now that Yoon has a surplus of frozen periodical cicadas, he can experiment in ways no one has before when preparing dishes of Brood X cicadas.

Under normal conditions, seventeen years is an inordinate, insufferable time to wait for a meal, so Yoon and others need to look beyond Brood X. With unmatched diversity, insects can easily fill a discerning gourmand's larder. A recent study confirms 1,611 species of insects have traditionally been consumed by humans, but the number is undoubtedly far higher. Truth be told, we all eat a variety of insects because bits of their bodies and their eggs invariably end up in our food. Title 21, Code of Federal Regulations, Part 110.110 allows the US Food and Drug Administration to set maximum levels of insects in everything from cinnamon to wheat flour (average of four hundred or more and seventy-five or more insect fragments per fifty grams, respectively). Insects are inside, on, and all around the plants we eat. Extract environmental DNA from a single tea bag, and you can find traces of as many as four hundred different kinds of insects, as Henrik Krehenwinkel and colleagues recently discovered from tea bags they purchased in Trier, Germany.

Technicalities aside, much of the world has, for ages, deliberately eaten insects for survival or as delicacies. Certain dishes featuring insects are clearly linked with cultures, and help identify and distinguish them. Peter Menzel and Faith D'Aluisio traveled the world to photograph and taste-test insects, resulting in *Man Eating Bugs: The Art and Science of Eating Insects*, a beautiful and anecdote-rich travelogue. Among manifold other culturally important

Cicada spring salad by entomophagy ambassador and passionate chef from Brooklyn Bugs, Joseph Yoon

and geographically specific insect dishes, they ate roasted witchetty grubs in Australia, sun-dried mopane worms in Botswana, and grasshopper tacos in Mexico.

Insect edibles have helped define our cultural identities for a long, long time. Like chimpanzees' and orangutans' use of tools to fish out and eat termites from termite nests, early hominids may have used antelope bones to excavate termites from their mounds. Archeologists debate about the use of ancient tools, and a reanalysis of 23,000 bone fragments from the Swartkrans cave site in South Africa

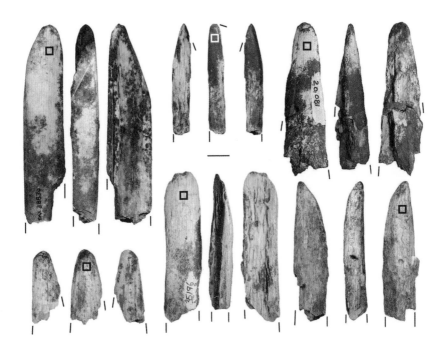

led scientists Lucinda Backwell and Francesco d'Errico to conclude that early hominids (*Homo* sp. or *Paranthropus robustus*) used a small subset of these bones to dig for termites. The patterns of wear on some ancient bones matched the wear on bones used by the researchers to excavate termite mounds, and this pattern was distinct from wear on bones used for digging soil, digging for tubers, tanning hides, or on bones of prey found in hyena dens. This means that a tradition of using bone tools to forage for termites may have existed in southern Africa for nearly one million years. Similar patterns and conclusions have since been found and drawn from bones in three additional cave sites in South Africa, resulting in a total of 102 bone tools, including those with markings indicative of termite mound excavation. Though contentious, bone tools offer additional evidence of our cultural reliance on insects.

Bone tools from the early hominid sites of Swartkrans and Drimolean in South Africa show the same wear pattern as found on contemporary bones used by researchers to excavate termite mounds. Lucinda Backwell and Francesco d'Errico made replicas of the wear patterns (marked by squares) and studied them microscopically.

Today, humans consume at least sixty-one species of termites, with a preference for the fatty immatures and winged reproductives of *Macrotermes*. Other great apes also eat termites, but which members of a termite colony they consume relates to their dietary needs. Chimpanzees supplement their protein-poor but micronutrient-rich fruity diet with the protein-packed soldiers of *Macrotermes*. Gorillas supplement their protein-rich but micronutrient-poor leafy diet with micronutrient-loaded *Cubitermes* workers. Julie Lesnik, biological anthropologist at Wayne State University in Detroit, Michigan, wanted to reconstruct the diets of our ancestors, and one approach was to compare our energy costs and dietary needs with those of our closest living relatives. She also used evidence locked in ancestral bones, in the form of isotopes (different forms of elements) to reconstruct a probable diet that further supports early hominid entomophagy.

If entomophagy served our ancestors so well, and our primate relatives and some human communities benefit from eating insects today, why are insects not a cosmopolitan staple across all human societies? Is the very thought of insects on your tongue detestable to you? Your emotional response will likely depend on the part of the world in which you feel most at home. Over one billion people intentionally eat insects, but finding insect-eaters in the United States, Canada, or most European countries is a challenge.

Lesnik notes in her book, *Edible Insects and Human Evolution*, that if you live in the tropics, you are far more likely to eat insects than if you live farther north or south of the equator. Latitude matters, but likely for a complex jumble of reasons, with reliability of insect abundance and spread of human colonialism top among them. Insects are always accessible in tropical regions, but are less abundant and accessible elsewhere, especially in regions with bitterly cold seasons. Relying on insects for sustenance in a temperate zone would be foolhardy, unless insects were gathered during warmer months and stored for leaner times, or were raised in climate-controlled settings.

Colonialism, an oppressive force that has affected almost all aspects of modern human culture, has also dramatically influenced global patterns of insect consumption. Christopher Columbus associated entomophagy with savagery, and such attitudes contributed to the dehumanizing and

mistreatment of people the world over. A Western-driven stigma against eating insects presents consequences, including affecting communities without good dietary alternatives to eating insects. Though Indigenous cultures within the United States had a rich history of eating insects, bias against entomophagy resulted in its decline. Documentation of entomophagy in North America is minimal, and this may be due to a reluctance by Indigenous communities to share details of their practices with prejudiced Westerners.

Edibles (2021) is a selection of insects and arachnids eaten by humans, painted actual size by Isabella Kirkland.

Another effect of latitude is farming, which, in contrast to entomophagy, has played a greater role as people settle farther from the tropics. The negative association between farming and entomophagy may stem from a view that insects are agents of destruction when it comes to cultivating crops. Insects can be viewed as a threat that spells loss of food, or food spoiling.

But another way to view these insects is as a prize rather than as a scourge. It is possible to capitalize on, rather than wither under the destructive weight of, insect infiltrations. If locusts ravage a crop, harvesting the locust may be an option. Weaver ants (*Oecophylla* spp.) are predators of insects that damage mango and cashew plantations, and can themselves be harvested as a sustainable food source for undernourished rural citizens, according to ecologist Joachim Offenberg and ant biologist Decha Wiwatwitaya. They eat insects considered pests by farmers of these crops, so act as agents to control agricultural pests and as a source of sustenance when properly managed or harvested.

Even when eating insects seems pragmatic, our dietary choices can be constrained by fear, or disgust. A reluctance to eat insects can come from associations of insects with disease or spoiled food, or the fear of eating something new. For some, the horror of recognizing animal parts in a dish can be off-putting. This can particularly affect those exclusively exposed to highly processed and packaged foods, and for whom seeing food for what it is can be a shock.

Reluctance to eating insects, however, can arise for many reasons beyond those relating to fear or disgust. Some people are restricted from consuming insects due to health reasons. The US Food and Drug Administration warns that if you are allergic to seafood, you may also be allergic to insects. This is because insects are relatively closely related to crabs, shrimp, and other commonly consumed (and allergenic) "shellfish."

Eating insects can also clash with personal principles. Many consumers may feel conflicted about eating insects as our understanding of what it means to be sentient or to feel pain extends to invertebrates. Insect welfare and the ethics of consuming insects are important topics to acknowledge as we grapple with our eating choices, today and in the future. If the treatment and slaughter of vertebrate livestock is discomfiting, shouldn't the killing of myriad times more insects pose similar discomfort? Should our ethical approach to eating depend on the numbers killed, the proximity to which we are related to those killed, or on some measure of their perceived worth, like beauty, ecological services performed, or their ability to learn or solve problems? For now, the ethics of eating insects falls to each consumer, who can weigh variables important to their personal life choices. For those guided by religious doctrine, some rites or scriptures attempt to clarify which insects, if any, are consumable. For Jains in India, killing insects is a karma-harming act of violence, so eating insects is forbidden. Jews abiding by biblical dietary laws can eat a select few insects, though what insect is kosher is up for (Hebrew) interpretation. The book of Leviticus (11:20-23) of the Old Testament, according to the Revised Standard Version (1952), states:

> All winged insects that go upon all fours are an abomination to you.
> Yet among the winged insects that go on all fours you may eat those which
> have legs above their feet, with which to leap on the earth. Of them you
> may eat: the locust according to its kind, the bald locust according to its
> kind, the cricket according to its kind, and the grasshopper according to
> its kind. But all other winged insects which have four feet are an abomina-
> tion to you.

Yet, the King James Version (1611), by including beetles and excluding crickets, exposes what may reflect a lack of familiarity with insects among the translators:

> All foules that creepe, going vpon all foure, shalbe an abomination vnto
> you. Yet these may ye eat, of euery flying creeping thing that goeth vpon
> all foure, which haue legges aboue their feet, to leape withall vpon the
> earth. Euen these of them ye may eate: the Locust, after his kinde, and the
> Bald-locust after his kinde, and the Beetle after his kinde, and the Grasse-
> hopper after his kinde. But al other flying creeping things which haue
> foure feet, shall be an abomination vnto you.

Adherents of the New Testament can look to Saint John, who ate locusts and wild honey for sustenance. Readers of the Koran will find mention of eight insects and a spider (depending on interpretation), and Islamic tradition, like the New Testament, cites the eating of locusts and honey.

Muhammad, in response to others using their whips and sandals to hit a swarm of locusts, is purported to have said, "Eat them for they are the game of the sea."

If eating animals is already part of your routine and you are not averse to eating insects out of fear, for health reasons, or based on spiritual or ethical principles, the environmental advantages are numerous. If you wish to minimize your impact on global warming, farming insects emits a lower amount of greenhouse gases per calorie than obtaining (in order, from most to least) beef, cultured beef, fungus protein, chicken, algae, pork, tofu, peas, jackfruit, and even beans. Far less water is used and far less waste results from insect farming than farming other livestock.

According to a champion of entomophagy, Arnold van Huis, for every ten pounds (4.5 kg) of food you feed cows you get one pound (0.45 kg) of beef. A 10 percent return on your biomass investment doesn't sound very good. Feed pigs, and you get two pounds (0.9 kg). Chickens, and you have four pounds (1.8 kg). Give ten pounds (4.5 kg) of food to crickets and your net profit is a more promising six pounds (2.7 kg) of "meat." How much of each animal can you eat? Forty percent of a cow, 55 percent of a pig or a chicken, and 80 percent of a cricket. Crickets are the most common insect livestock reared today, though other insects are being farmed around the world. Soldier fly larvae can match or exceed the net profit of crickets, but whether you can acquire one pound (0.45 kg) or seven pounds (3 kg) of edible material after feeding the larvae ten pounds (4.5 kg) of food depends almost entirely on what you feed them.

Facilities to mass rear crickets or other insects for their high protein yields are cropping up. Recently completed, the Madagascar Biodiversity Center is a research and insect-rearing facility with aspirations to reduce Madagascar's habitat destruction and loss of biodiversity. Many questions remain about how best to shift our production to insect-based food, and animal (including insect) welfare is one of many ethical and health-centered discussions that will shape how future generations treat animals slated for the chopping block, as well as which animals would do the least environmental harm and the most economic good.

But eating insects need not constitute a personal sacrifice to save the planet. Your reasons can be purely nutritional or gastronomical. Like Roald Dahl's short story "Royal Jelly," in which a beekeeper secretly slips royal jelly produced by his honey bees into his malnourished baby's bottled milk, many turn to insects for nutritional gains. What will really win over skeptical consumers, however, is that insects appear appetizing. Vincent Holt, in his 1885 appeal to human reason, *Why Not Eat Insects?*, wrote "While I am confident that they will never condescend to eat *us*, I am equally confident that, on finding out how good they are, we shall someday right gladly cook and eat *them*." Joseph Yoon, our cemetery harvester and passionate advocate for entomophagy, is one of a growing number of chefs incorporating insects into dishes with the primary aim of showcasing insect delicacies to please the palate.

Personal Entomophagy Sampler

Hans Bethe, Nobel Prize–winning physicist and former head of the theoretical division of the Manhattan Project, let me enter his Cornell University office—what looked to me like a cluttered bunker lost in time. Remaining in his seat behind his desk, he waited for me, an undergraduate student, to state my business. I felt like an intruder and began to second guess my intentions. What could possibly be worth interrupting the work of someone who Freeman Dyson called the "supreme problem-solver of the twentieth century"? I wondered this very thing as I proceeded to invite him to an event I was staging at a campus cooperative—a dinner with insects. In the middle of my rattling, Dr. Bethe slowly lifted his ancient arm, signaling me to pause. Then, with disconcertingly slow motions he . . . opened . . . his . . . desk drawer, pulled . . . out . . . and . . . inserted . . . a hearing aid in his ear. After this prolonged pause, he blankly stared, waiting for me to start over. So I did. After he realized I was not there to discuss stellar nucleosynthesis or solid-state physics, he simply said, "No." I thanked him for his time, then slinked out.

My visit to invite Carl Sagan, astronomer, author, and television host of *Cosmos*, went much better. I entered, for the first time, the Space Sciences Building and tried to find Sagan listed in the directory posted on the wall. No one by that name. I asked an administrator and she sighed, responding that they no longer posted his name because it had been

Entomophagy by
Barrett Klein

stolen too many times. She directed me to his office (room 302), where I found, seated in a surprisingly small space at a tiny desk, one of the most effective communicators of science to have ever lived. He jovially invited me in, listened to my insect-eating overture with delight, and posted my invitation (featuring a drawing of an insect composed of vegetables) on his office wall. He kindly let me know that he may not be able to attend, but thanked me repeatedly, with grace and sincerity. Neither Bethe nor Sagan showed up, but the event introduced those in attendance to the cultural practices and history of entomophagy, with a menu featuring cricket jambalaya, mealworm salad, whole wheat bread supplemented with mealworm flour, pollen cookies with honey, and cochineal-dyed beverages, plus a cameo appearance by silkworm moth pupae. More elaborate events with a far greater diversity of insects exist, but this was a somewhat unusual occasion on a college campus in 1993.

Eating insects doesn't always require the planning of cultural enrichment programs or delivery of vegetable-insect invitations. As a master's student at the University of Arizona, I returned from class to find that the honey jar I had left on my counter was not tightly sealed, and this had allowed a stream of tiny ants to form a line to, up, and into the jar. A layer of unsuccessful foragers lay dead on the surface of the honey, and this number was growing because those who made it out alive were laying a chemical trail to recruit additional waves of nestmates. Not one to waste food, I slathered some honey on bread, ate it, and found, to my surprise, that it tasted remarkably like blue cheese. To check whether the honey or bread had gone bad, I tasted each in turn. Normal. Then the ant-filled honey. I was far from the first person to discover that a crushed odorous house ant (*Tapinoma sessile*) smells (and, in my case, tastes) like blue cheese, though some liken the smell to rotten coconuts, rancid butter, or cleaning solution. On a mission to uncover the chemical culprit behind the odor, Clint Penick and Adrian Smith pinpointed a methyl ketone (6-methyl-5-hepten-2-one), noting that methyl ketones give blue cheese its distinctive odor. Other ants, when crushed, smell of lemon or vinegar, and these additional sensory cues can help us identify the ants. They can also add to their palatability.

Honey is one example of a soft approach to entomophagy. Instead of eating insects, we can eat what they regurgitate. A potentially unwise example with delectable consequences took place in Panama, where I had just cooked a remarkably unremarkable plate of pasta. A large katydid flew through my window, landed on my plate, and regurgitated copious amounts of defensive chemicals on my noodles as I attempted to lift her away. Though I had never experienced such a katydid expulsion, I knew from experience that, every now and then, the defensive regurgitation of grasshoppers could be uniquely spicy. Noting the culinary potential, I coaxed her to provide more sauce before setting her free and, as a result, enjoyed a much more flavorful meal.

Why might her chemical deposit be tasty to some palates? If you look closely at many of our spicy or bitter foods, the origin of the flavors we covet stems from a plant's response to being attacked. Spicy peppers rely on capsaicin to activate a protein we normally use to detect heat. Peppers evolved this chemical trickery because those affected most are herbivores who tend to avoid eating them, whereas birds—their seed dispersers—are unreceptive and unfazed by capsaicin. Mustard is a staple for many humans, yet its piquant nature evolved in response to hungry insects, repelled by the spicy quality we often seek. Regional variations in mustard flavor appear to be driven by localized insects' appetites. A ceaseless arms race between insects and plants has produced the chemistry of so many flavors and floral fragrances that tantalize our taste buds and entice and tempt us through our sense of smell.

The katydid serving as my sous chef was simply regurgitating her latest meal, which was a plant attempting to chemically dissuade herbivores from doing what they do, and co-opted by the katydid to dissuade me from consuming her.

Many insects exude chemical defenses. Although licking the occasional stinkbug's glandular secretions or grasshopper's "tobacco juice" can be uniquely flavorful, occasionally offering a cinnamon-like zing, I cannot, in good conscience, recommend indiscriminately licking your backyard insects. One example of a taste test gone bad: I picked up a flamboyantly colored moth in Costa Rica (pink! orange! white!) and the moth promptly left a small pool of brown regurgitant on my palm. I dipped my tongue in, and dedicated the next portion of my day attempting to extract the harshly bitter taste from my mouth.

Medicine

What, if anything, would compel you to incorporate insects into your diet? We touched on how eating insects plays a role in traditions and customs that help identify cultures, how as an alternative protein source insects can help with food security and global sustainability, and entomophagy provides nutritional benefits, as well as novel culinary experiences with unique flavor profiles. There is yet another reason people have eaten insects. When carefully selected or prepared, certain insects can provide medical benefits. It is important to take the qualifier "carefully selected or prepared" seriously. Just as we've seen examples in which people (including the author) throw caution or rational thought to the wind when communing with insects, there are clear examples of deception or ignorance when it comes to the medical efficacy of entomotherapy (see Honey, page 74, for examples).

Ancient medical texts often attribute lofty healing powers to humble ingredients. Is it possible that eating solitary wasps (*Ammophila infesta*) is effective against long-standing deafness, cough, and bad breath, as Pen Ching reported long ago? (Useful aside: The same wasps, according to the "Kou Lou record," "should be collected on the fifth day of the fifth moon, dried in the shade and powdered, and made into pills with human blood from a man slain in battle, and placed in the collar. It will make everyone afraid of the person so treated.")

The historical applications of insects for healing are endless and some are preposterous, but to indiscriminately dismiss traditional medical practices would be a tragic mistake. Folk medicines are still one of the most important paths for discovering drug resources from nature. From nearly every part of the world, oral traditions tell of insect treatments, and some of these are millennia old. Many of these treatments involve using insect products (honey, propolis, beeswax, silk, nest material, egg cases, shellac—for knife wounds!—and venoms) or fungus-infected insects (*Cordyceps* sprouting out of caterpillars), but here we will think about examples in which insects themselves are used for healing purposes. Eraldo Medeiros Costa-Neto, when reviewing the use of insects as folk medicines in the Brazilian state of Bahia in 2002, reported a great variety of approaches when preparing insects to treat ailments. Insects are crushed, fried, boiled, toasted, and powdered, or squeezed directly onto afflicted areas. Most are administered as teas. If a child has asthma in Matinha dos Pretos, a matchbox containing a live black cockroach is slowly opened close to the child's nose three times per day. Will this controlled exposure help boost the child's immune system? Costa-Neto documented that at least forty-two insect species are used in folk medicines in Bahia alone.

No one knows the total of how many insects have been used medicinally, and how many of these folk remedies are effective will remain utter speculation unless more concerted efforts are made to rigorously test them. Connecting folk medicine to modern treatments has great potential, given their sustained use in many cultures, and seeing as insects have evolved the means of synthesizing or sequestering a profuse array of bioactive chemicals. Insects are chock-full of glands producing potent chemicals for communication, antimicrobial protection, defense against predators, or to subdue, paralyze, control, or kill. As we saw with venoms and poisons, this can come in handy in surprising ways . . . for us. Extracts taken from a variety of insects show anticancer activity, and compounds derived from chitin (found in insect exoskeletons) are used in drug delivery systems, and to dress wounds.

As I write this, the world struggles with what may continue to be an ongoing string of pandemics, and COVID-19 remains a threat, despite the availability of a battery of vaccines to combat it. Developing more than one vaccine to battle a virus has multiple benefits, including offering people allergic to an ingredient in one vaccine alternatives so they can still acquire protection against the virus. For example, polyethylene glycol, an ingredient found in mRNA vaccines (Moderna and Pfizer/BioNTech), can cause anaphylaxis in some patients, so an alternative is essential for them.

Novavax is a COVID-19 vaccine that does not include polyethylene glycol. Instead, Novavax includes moths. Well, it includes proteins created by cells from moths—the fall armyworm, *Spodoptera frugiperda* to be precise. What researchers have figured out is that you can take genes known to code for proteins that trigger an immune response, insert these genes into a virus that attacks insect (fall armyworm) cells to mass produce these proteins, then collect and purify the proteins. To amplify the response, they add an extract from soapbark trees. COVID-19 is one of the deadliest pandemics in human history and some of us owe our protection to moths and the researchers manipulating their cellular machinery.

Sometimes the solution to a malady requires a far cruder approach than genetically engineering viruses that co-opt moth machinery. Say you're bushwhacking in a South American tropical rainforest and, falling, tear your skin against one of the many plants bearing protective thorns. A gaping wound needs to be closed, but access to surgical equipment may be too far away to be of help. Fortunately for you, a traditional source of sutures frequently saunters across the forest floor. Army ants. Legions of worker ants scour the forest for any animal that cannot escape the grasp of their primary weapons: mandibles, or jaws, that clasp. A specialized caste of soldiers has such large sickle-shaped mandibles that they can use them only for biting, not for eating. A pair of long, sharp, curved mandibles clamps and remains locked. It is easy to see how people from different parts of the world converged on a similar solution of using the formidable jaws of still-living ants, termites, or beetles to close wounds. Folk healers in the Balkans, Algeria, and Zambia used insect mandibles as sutures, and several fourteenth- through sixteenth-century Italian surgeons used ant mandibles to close wounds in the small intestine. A description from 1925 details the approach taken using *Camponotus* carpenter ants by the de facto surgeons of Smyrna, the barbers of Greece:

> *The barber presses together the lips of the wound with his left hand, and applies each ant by means of forceps held in his right hand. The mandibles of the ant being wide open and the animal in a defensive attitude, when*

the insect is slowly brought to the wound it seizes the outstanding surfaces
as soon as it has been brought to them, sinks its mandibles into the flesh
on both sides of the wound, and remains in this position, closing each
mandible against the other vigorously, and consequently holding the two
edges tightly to each other. Then the barber separates the head from the
thorax by a snip of the scissors, and the head with its mandibles remains
in place, continuing its office though the body has fallen to the ground.

The process is repeated, head after head, and these remain in place for several days until "the heads are removed, their office being no longer necessary." With evidence showing that a honey bee's bite (not sting) comes with a local anesthetic, it's possible that traditional mandibular sutures may come with additional, not-yet-confirmed, benefits.

I will close this foray into the healing arts with a practice some might assume I dredged up from either a medieval manual on torture or a misguided guide to folk medicine, discarded long ago in the name of common sense and common decency, and rendered obsolete thanks to medical advances. First, picture walking a dog. As long as weather permits, the inevitable deposition of dung by your dog results in the equally inevitable visit by flies to that dung. Ignore your impulse to be a good neighbor and leave the faex in place. Return and you will find fly larvae—maggots—feeding on what your dog has left. The same will happen with roadkill, or garbage left to rot. Now picture purposefully placing the same fly larvae that would develop in dung or garbage on a wound so that the larvae can consume the flesh of the wounded. If your wound is oozing, moist, and exposed to enough oxygen, you are a candidate for maggot therapy.

Traditional healers and modern medical practitioners alike deploy the maggots (for example, the common green bottle fly, *Lucilia sericata*) for two reasons: No surgeon is better equipped to remove necrotic tissue from a wound than maggots. Within two to three days, the larvae will liquify and ingest all the dead tissue, including tissue hidden from view. When they do so, they also disinfect the area. It is unsanitary to eat in a pathogen-ridden zone, so the flies have evolved a chemical means of keeping their environment

Lucilia sericata, the green bottle fly with a penchant for dung, garbage, and carcasses, can save human lives in their larval form.

free of disease. Anecdotal evidence, including evidence gathered from Civil War battlefields, indicates that patients with fly-infested wounds experience reduced fevers and are more likely to survive than those deprived of flies. The Maya, Native Americans, Aboriginal tribes in Australia, and Europeans have all turned to maggots to assist in healing. This is not an extinct procedure. The stigma associated with using maggots fades with each gangrenous limb saved from amputation, and wider medical use of the flies may be on the horizon. If you suffer from diabetic foot ulcers, flesh-eating bacteria, or other nonhealing wounds, you would be wise to welcome maggots into your (professionally supervised) treatment regimen.

Light

Vision requires light, and in the darkest recesses, the deepest oceans, or when the sacred scarab beetle has rolled the sun below the horizon, the only natural light comes from other celestial bodies, or the bioluminescing bodies of some bacteria, dinoflagellates, mushrooms, or animals. If alive, why glow? Some ocean dwellers use light to camouflage themselves by matching their illumination with skylight above. Predators on the hunt below see no dark body contrasted against the sky and pass up the counterilluminated meal. Some species of squid expel clouds that glow to distract or repel. For others, attracting predators can be a boon. If captured by one predator, flashing light can attract a second predator, setting up a kerfuffle and an opportunity for escape. Or maybe it pays to be eaten, so glow and a predator can spread your spores more effectively than you can alone. Changing vantage points, some predators also have reason to emit light. A deep-sea angler fish dangles and flicks a lure full of bioluminescing bacteria to attract prey, while deep-sea barbeled dragonfishes shine a red spotlight to locate prey. Long, red wavelengths have the least energy to penetrate water, so there is normally no red light in the abyss, so creating your own red light can offer the benefit of clandestine communication, or spotting organisms incapable of perceiving that rarest of hues at the bottom of the sea.

The most celebrated among the illuminators are the fireflies, the lightning bugs—neither flies nor true bugs, but beetles. Fireflies emit light by combining luciferin and luciferase (from lucifer, "light-bringer") in the presence of other chemicals, like oxygen, and erupt with species-specific displays to attract mates. Courtship can be risky, and there are femme fatale fireflies (*Photuris* spp.)

A Visual Compendium of Glowing Creatures (2014), Eleanor Lutz's chart of bioluminescent species and the chemical structures allowing them to glow

THE BIOLUMINESCENT TREE OF LIFE

DINOFLAGELLATE
Lingulodinium polyedrum
10 μm

SEA SPARKLE
1 mm
Noctiluca scintillans

GHOST FUNGUS
10 cm
Omphalotus nidiformis

SOUTH AMERICAN LEAF SPOT
1 mm
Mycena citricolor

HONEY MUSHROOM
10 cm
Armillaria mellea

SEA WALNUT
1 cm
Mnemiopsis leidyi

CRYSTAL JELLY
1 cm
Aequorea victoria (Aequorea aequorea)

COLONIAL RADIOLARIAN
1 mm
Collozoum inerme

PROTISTS
Other eukaryotes

BITTER OYSTER MUSHROOM
5 mm
Panellus stipticus

MOLDS & MUSHROOMS
Fungi

COMB JELLIES
Ctenophora

OVOID COMB JELLY
1 cm
Beroe ovata

COLONIAL HYDROID
1 mm
Obelia longissima

CHO
1 μm
Photobacterium leiognathi

CHO
1 μm
Aliivibrio fischeri (Vibrio fischeri)

CNIDARIANS
Cnidaria

HYDROZOANS
Hydrozoa

CROSS JELLYFISH
1 cm
Mitrocoma cellularia

COLONIAL TUNICATE
5 cm
Pyrosoma atlanticum

BACTERIA
Bacteria

CORALS & SEA ANEMONES
Anthozoa

JELLYFISH
Scyphozoa

HELMET JELLYFISH
10 cm
Periphylla periphylla

SMALL LANTERN FISH
1 cm
Diaphus holti

CHORDATES
Chordata

ACORN WORMS
Hemichordata

SEA URCHINS & STARFISH
Echinodermata

SEA FEATHER
10 cm
Ptilosarcus gurneyi

SEA PANSY
1 cm
Renilla reniformis

ALARM JELLYFISH
5 cm
Atolla wyvillei

HATCHET FISH
1 cm
Argyropelecus hemigymnus

ACORN WORM
1 cm
Ptychodera flava

BRITTLE STAR
5 mm
Ophiopsila californica

PROTOSTOMES
Protostomia

WORMS & LEECHES
Annelida

PARCHMENT WORM
1 cm
Chaetopterus variopedatus

ASSISTED BIOLUMINESCENCE
Most shallow water glowing fishes use other bioluminescent species to glow

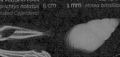

CHO
LUMINOUS EARTHWORM
5 cm
Diplocardia longa

POLYNOID SCALEWORM
5 mm
Harmothoe lunulata

STOPLIGHT LOOSEJAW
1 cm
Malacosteus niger

EYELIGHT FISH
1 cm
Photoblepharon palpebratus (symbiotic bacteria)

SWEEPER FISH
1 cm
Parapriacanthus beryciformes (ingested Cypridina)

MOLLUSCS
Mollusca

ARTHROPODS
Arthropoda

PLAINFIN MIDSHIPMAN
5 cm
Porichtys notatus (ingested Cypridina)

SEA SNAIL
1 mm
Hinea brasiliana

COMMON PIDDOCK
1 cm
Pholas dactylus

SNAILS & SLUGS
Gastropoda

FIRE CENTIPEDE
1 mm
Orphnaeus brevilabiatus

GENJI FIREFLY
5 mm
Luciola cruciata

CENTIPEDES & MILLIPEDES
Myriapoda

INSECTS
Insecta

SOUTH AMERICAN RAILROAD WORM
1 cm
Phrixothrix hirtus

FRESHWATER LIMPET
1 cm
OCH₃
Latia neritoides
CHOCHO

FIREFLY SQUID
1 cm
Watasenia scintillans

SQUID, CUTTLEFISH, OCTOPUS
Cephalopoda

CRUSTACEANS
Crustacea

FLYING SQUID
5 cm
Symplectoteuthis luminosa

GLOWING SUCKER OCTOPUS
5 cm
Stauroteuthis syrtensis

DEEP SEA SHRIMP
1 cm
Sergestes similis

COPEPOD
100 μm
Metridia lucens

SEA FIREFLY
Vargula hilgendorfii (Cypridina hilgendorfii)

SIERRA LUMINOUS MILLIPEDE
5 mm
Motyxia sequoiae

KRILL
5 mm
Euphausia pacifica

1 mm

A chart of selected bioluminescent species and associated bioluminescent luciferins, luciferases, and photoprotein structures. Based on the textbook *Bioluminescence: Chemical Principles and Methods* (Revised Edition) by Dr. Osamu Shimomura.

2019 Eleanor Lutz

who mimic the light displays of females belonging to other firefly species (*Photinus* spp.). This mimicry attracts male *Photinus* fireflies, and the unsuspecting suitors are lured into the waiting clutches of a hungry predator. The femme fatale fireflies benefit from more than the nourishment of a duped male. *Photinus* produces defensive chemicals (lucibufagins) that *Photuris* cannot, so the more *Photinus* a *Photuris* female consumes, the more repulsive she is to her own predators.

Fireflies' nocturnal glow is depicted in countless pieces of art, films, and animations, and treated in songs and stories. Their allure is most jubilantly on display at firefly festivals in pockets of the United States (Pennsylvania, New Jersey, Virginia, South Carolina), and in Purushwadi, India, and, most notably, Japan. Though often romanticized, fireflies have also been harnessed for their utility. Less obvious or glamorous applications in Japan include fireflies in drugs, ointments, and as woodworkers' grease. Where fireflies shine brightest, however, is when that luciferin-luciferase connection is on display. The molecular process operating in fireflies to produce the glow has been co-opted for medical technologies, and has taught us more about the mysterious human, Neanderthals, with whom our ancestors interacted. Neanderthals have been genetically sequenced, and lessons we can extract from their DNA is, in part, because of lightning beetle biochemistry.

It's difficult to imagine during our post-industrial time, awash in light pollution, that people used to be light-limited. Activities had to alter or cease radically after sunset due to a scarcity of photons. Fireflies, at times, were caged light-bringers for those wishing to extend their day. Deirdre Prischmann-Voldseth, in her survey of art with fireflies, uncovered three such references. One tells the story of a Chinese student who imprisoned fireflies in a paper lantern to study after dark, later to become an eminent scholar. The other two are works of art, also involving the light of fireflies to facilitate reading, the first on a fan painted by Sha Fu (1879), and the second on a knife handle by Yokoya Sômin V (mid-nineteenth century).

A firefly lamp hangs from a branch and lights scrolls in Sha Fu's *The Peach and Plum Garden* (1879; detail of folding fan, ink and color on alum paper).

Insect as Guide

I don't need to stretch my imagination too far to appreciate the utility of an insect's light because there was a time when I depended on one to safeguard me from danger. The place: La Selva Biological Station in Costa Rica. I was curious about the strange behavior of tiny ants that ride as hitchhikers atop leaf fragments, cut by their nestmates from a tree canopy and carried all the way down by these much larger workers to feed their subterranean gardens of fungi. These farmers predate human agriculture by sixty million years. I came to study them.

Early one evening, David Wagner—entomologist extraordinaire—introduced me to the practice of attracting click beetles (*Pyrophorus* spp., family Elateridae) with light. If you rapidly move a flashlight in a downstroke then promptly cover the light, he explained, you can lure in a bioluminescent click beetle, their glowing patches resembling headlights. I found this hard to believe, but after he left, I followed his instructions using my headlamp and after a grand total of one attempt . . . thump! . . . a hefty beetle struck my chest. I placed it in a clear vial with the intent of showing and sharing it with others later that night. First, I wanted to take a solo hike into the rainforest.

The sensory experience of a rainforest at night is exhilarating. A cacophony of courting frogs and insects, aroma of fruits and wet earth, and humidity on the skin can be more easily appreciated at night, when

A bioluminescing click beetle (*Pyrophorus* sp.) from Costa Rica controls the intensity of light constantly emitted from a pair of yellow spots on the prothorax. They also emit light from an organ on the anterior underside of the abdomen.

visibility is diminished and the diurnal cast of characters has made way for their nocturnal counterparts. To make my way along the narrow trail under the dark canopy, I depended on the light of my headlamp. I stumbled upon a thick trail of leaf-cutter ants running up and down a tree, carrying leaf fragments laden with hitchhikers. The temptation was too great. I pulled out my notebook and collected data. At some point during my fit of spontaneous science, I noticed that my lamp's light was starting to dim. I did the obvious thing. I hastened the pace of data collection.

The light eventually faded to nothing and I was left in total darkness at least one kilometer from the station. This forest is noted for its aggressive and venomous fer-de-lance snakes, among other night-active organisms, so carelessly stumbling back would have been unwise. Instead, I pulled out my vial with the bioluminescing click beetle, held it close to the ground, and ever so slowly made my way back to the station, where I could show those still awake the marvel of my glowing guide. Only later did I learn about sixteenth-century Islanders in the West Indies lighting their paths by tying these beetles to their toes.

Art

Your job is to move grains of sand. Clear them out, make way for others as you dig. Tunnels form, allowing your nestmates to travel, find protection, and expand their reach. Your job is simple. You are motivated, though no leader commands you to continue your digging. Focused and fixated, you are distracted only when you pass nestmates and shuffle your antennae with theirs to confirm allegiance and rule out foreign invaders. Were you to step back and view the actions of your entire colony, the frenetic process would appear utterly unfamiliar and discombobulating. A global view of your collective efforts underground is never visible to any inhabitant, and normally such a view would be impossible for anyone else to see. But you are hauling sand in the middle of a brightly lit museum behind clear plastic walls. One transparent, wall-mounted plastic box is connected to another by a transparent tube, and that one is connected to another, and another. You are one of thousands of ants building a nest within a collection of boxes full of colorful sands, precisely sprinkled by a visionary unknown to you, an artist who created the flags of the world from the sands you and your nestmates are gradually mixing. Humans peer at your studious work as they see a larger picture imposed upon you. The flags represent arbitrary political boundaries of which you know nothing, and do not respect. They are becoming less and less recognizable as your work continues and your nest takes shape. What seems so very important to one species on the planet means absolutely nothing to you, as you go about your way creating your own world.

The unseen hand was that of Yukinori Yanagi, an artist whose political statements have largely been conveyed through the activities of harvester ants (*Pogonomyrmex* sp.). When Yanagi prepares *(continued on page 161)*

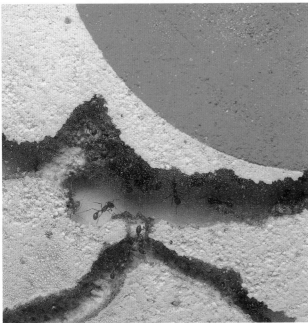

ABOVE: Flags of the Americas, composed of colored sand in plastic boxes. The sand is shuffled around by a colony of ants, breaking down symbols of political separateness in Yukinori Yanagi's *America*.

LEFT: The close up shows Yanagi's harvester ants in action, this time dispersing sand forming Japan's flag.

Guess which works of art include insects or insect bodily products!

(Answers on the following page.)

Key:

A. Yes! Nick Knight and Alexander McQueen arranged eighty gallons of live maggots; the fly larvae were color-coded, having eaten foods dyed different colors.

B. Yes! Suze Woolf created an expandable book from this tree branch covered with galleries burrowed by bark beetles in *Bark Beetle Book Vol. XXXVI: The Sky Cracks Open* (2021).

C. Yes! To create *Starry Night* (2004), Steven Kutcher daubed the feet of hissing cockroaches (*Gromphadorhina portentosa*) with paint and turned the canvas on a lazy Susan as the insects walked.

D. Yes! A common practice, Augustus Saint-Gaudens coated this plaster bust of *Davida Johnson Clark* with shellac from lac insects (1886).

E. Yes! E. A. Séguy designed a portfolio of pochoir *Insectes* (1920) with arrangements of insects or insect designs.

F. Uh, yes. That's Albrecht Dürer's *Stag Beetle* (1505, watercolor and gouache).

G. Yes! The vulture and cobra on Tutankhamun's funerary mask were cast using the lost wax process, and the beard has been reattached with beeswax, a common material used during his time (circa 1323 BCE, Egyptian Museum).

H. Yes! The representation of a lightning beetle floats above others in this Mayan vessel with mythological scene (attributed to the Metropolitan Painter, seventh to eighth century CE).

I. Yes! Edgar Degas used pigmented beeswax (and other ingredients) for the original sculpture, and the bronze of *The Little Fourteen-Year-Old Dancer* (1922), cast after his death, wears a ribbon made of silk.

J. Yes! Petrus Christus's *Portrait of a Carthusian* features a fake fly resting on the fake frame (1446).

K. *Winged Victory of Samothrace* has no obvious insect connections (marble; circa 200–190 BCE).

L. Yes! Marlène Huissoud covered silkworm moth cocoons with honey bee propolis for *Cocoon Wardrobe* (2017).

M. Yes! This unlined summer kimono is made of silk and features insect cages, crickets, and grasshoppers.

N. *Mangaaka Power Figure (Nkisi N'Kondi)* has no obvious insect connections (Kongo artist and nganga, Yombe group, nineteenth century).

O. Yes, times three! This small detail of a festival banner about Krishna rescuing and marrying Rukmini is embroidered with silk and dyed red with carmine (cochineal) and lac (circa 1800, India).

P. Yes! *Arrangement in Grey and Black No. 1 (Portrait of the Artist's Mother*, 1871) is signed by James Abbott McNeill Whistler with his characteristic butterfly (on curtain, to left of picture on wall).

a work of ant art, he has to consider the biology of his subjects. Ants can only come from a single colony, otherwise there would be infighting. He chooses not to include a queen, so that the colony does not continue to reproduce beyond the capacity or longevity of the piece. He must feed them, so custom-made boxes with food are secured to the backs of the displays. Working with living insects, like Yanagi's ants, adds real-time action and drama to a piece, but as we've seen before, there are a variety of ways an artist can harness insects to express emotions, forward ideas, or highlight problems in the world. What follows are examples of what it means to create, what I like to call, "insect media art."

You can dress up to look like a beetle (that's coming later), and you can hang a picture of a butterfly on your wall, but these are not the same as displaying actual insects as art. Insect media art is any piece of art that includes an insect bodily product, an insect body part, or an entire insect body (live or dead). We've already visited insect media art when considering human uses of insect products, like silk, cochineal, shellac, and beeswax. These are the somewhat secret ingredients spanning the history of art. A small fraction of the artists who have worked with insect media, and an even smaller fraction of those who view their art, are aware that insects play a key role in the art's creation. And, of course, there are accidents. A grasshopper became mired in oil paint on one of Vincent van Gogh's canvases, and the wee remains were only discovered 128 years later by an observant conservator in Kansas City, Missouri. Calling *Olive Trees* (1889) insect media art, however, might be like calling your inadvertent consumption of fruit fly eggs when drinking a glass of orange juice an act of entomophagy. What is far more compelling than accidents is when artists use insects intentionally.

From vibrant traditional rituals to fashionable installations of contemporary art, insects, or key parts of their bodies, take center stage more often than one might expect. The prized component of some insects is their wings—structures that give insects not only flight, but can provide physical protection, camouflage or warning coloration, and, sometimes, sex appeal. Structurally, esthetically, or symbolically, wings are often the feature of choice to be embroidered into textiles, strung as jewelry, or glued to canvases or cultural objects of interest. Beetle wings, at least the front pair, are famous for their durability and beauty. Called "elytra," this front pair of wings serves as sheaths, used to protect the softer, membranous hindwings folded

Aguaruna ear ornament, with elytra (forewings) from metallic boring beetles (family Buprestidae), and feathers

underneath when a beetle is not in flight (Coleoptera, the order of beetles, is derived from Greek words meaning "sheath-wing"). Elytra adorn Indian paintings and attire from the Americas (Amazonia, Central America, and Mexico) to Australia and New Guinea. Young, unmarried Pwo Karen women in Myanmar and Thailand wear "singing shawls," with a tinkling, jingling fringe of elytra, beads, coins, and buttons, at funerals. The shawls produce a sound believed to ward off evil and escort the dead to the afterlife.

The Hall of Mirrors within the Royal Palace in Brussels, Belgium, with roughly 1.4 million elytra from aptly named jewel beetles (*Sternocera aequisignata*) shimmering from ceiling and chandelier (*Heaven of Delight* by Jan Fabre, 2002)

Some elytra are colorful to our eyes because of pigments that, with time, will break down and fade if displayed in the sunlight. Others appear colorful because of a structured surface through which light bends. Refraction of light through these structures gives beetles a magnificent iridescence that will glisten long after this book is ravaged by termites and silverfish and its contents are forgotten.

Far more delicate are the wings of butterflies, yet these are the stuff of collages by Jean Dubuffet, Damien Hirst, and others. Insect wings flitter here and there in different artists' works, but I know of only one person who has created still lifes using individual scales from butterfly wings. Insects can be hairy (though only mammals have true hair), and some lineages of insects can also be scaly. The order of moths and butterflies (Lepidoptera, derived from Greek words meaning "scale-wing") are beloved for their colorful wing scales, each scale with a smooth underside, and intricately structured upper side that scatters light to our eyes. Henry (later Harold) Dalton was a Victorian microscopist who skillfully assembled as many as one thousand butterfly scales, along with diatoms (unicellular, photosynthetic, ocean-dwelling "jewels of the sea") on a single microscope slide. He constructed microscopic still lifes with the scales and diatoms using a boar's bristle, a breathing tube, and only the internal oil of a wing's scale as adhesive. Dalton's process is described (slightly modified) here by The Museum of Jurassic Technology, a Los Angeles–based museum that revels in the spirit of wonder, and displays eight of Dalton's surviving slides:

After devising a design, Dalton would collect numerous butterfly wings of multiple species from all over the world. Carefully stripping off individual scales with a needle, each scale was then sorted by color, size, and shape, creating an extensive palette. Boar bristle in hand, Dalton would then transfer each scale to the slide. Positioning a scale was a laborious task, one that required the use of a microscope and a small tube through which he would breathe to gently move each scale over the glass to its appointed position. Once in place, Dalton would crush a tiny spot of the scale against the slide, allowing internal oils to act as a natural adhesive.

Artists have resourcefully collected and imaginatively painted, rearranged, assembled, mechanically animated, transformed, and dramatically displayed other insect parts, from head to tarsus (foot), or incorporated entire insects in their work. From the simple to the monumental and elaborate, insect media art is found in villages, on urban streets, and in the full gamut of museums.

Horns of rhinoceros beetles and the legs of ants or beetles are strung onto fibers to form jewelry, as is tradition among certain communities in South

Mosaics composed of diatoms and butterfly wing scales, assembled on microscope slides (3 × 1 in. / 7.6 × 2.5 cm) in the late nineteenth century by Henry Dalton

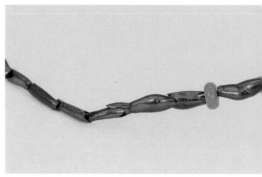

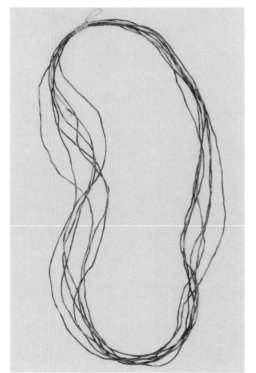

TOP LEFT: Heads: Strung on tucum fiber (*Astrocaryum arenarium*), horned scarab beetle heads (Dynastinae) are spaced with small black beads to form a necklace created by members of the Halót'ésú band of Nambikwara in Mato Grosso, headwaters of Rio Juina, Serra Azul, Brazil.

TOP RIGHT: Leg segments: Necklace strung with iridescent green beetle leg segments, from Matema Island in the Bismarck Archipelago, Papua New Guinea

BOTTOM: Leg segments: Strung on tucum fiber (*Astrocaryum arenarium*), ant leg segments form the beads of a necklace created by members of the Niyahlósú band of Nambikwara in Mato Grosso, Rio Camararé, Aldeia Camararé, Brazil.

ABOVE: Bodies: The encysted bodies of immature scale insects (*Margarodes* sp.), strung into a necklace by Zulu people of southern Africa

LEFT: Bodies: A band of scarab beetles line the base of a man's headdress from western New Guinea highlands. The green and beige beetles are facedown, set on red cloth, and visible through plaited orchid vine fiber. Above the beetles is a hat of human hair, topped with a central tuft of cassowary feathers.

America or Papua New Guinea. Immature scale insects (*Margarodes* spp.), dug from the soil, and (misleadingly) called "ant eggs" or "ground pearls," serve as lustrous beads in necklaces made by the Zulu people of southern Africa. And on the urban streets of London . . . a bumblebee is shot dead by a man standing by a girl clutching her teddy bear in *They're Not Pets, Susan* (2007) (see page 306), one of many miniature, staged dramas made with modified train set figures (and, in this case, real bumblebee) by street artist Slinkachu. A fellow British street artist, the anonymous Banksy, snuck an insect creation into the American Museum of Natural History in New York City. After adding sidewinder missiles and a radio dish to a harlequin beetle placed in a shadow-box (*Withus Oragainstus*, 2005), Banksy covertly affixed the piece to a column standing between a forest diorama and case of endangered species in the muse-um's Hall of Biodiversity. The stunt was brief before it was spotted, removed,

and stored among the museum's archives, where it has been coveted by museum archivists ever since.

Another way to display insects as art is to use their bodies as canvases. Catherine Chalmers painted live cockroaches to see if our perceptions of cockroaches could change if they look more like flowers or ladybugs (*Impostors* series). Akihiro Higuchi paints meticulous designs on the backs of dead beetles, leaf insects, moths, mantids, and others.

Some artists rearrange insect body parts to form fantastical creatures. Others animate their remains. The idea of animating insect specimens to appear alive harkens to the dawn of stop-motion animation, the filmmaking technique that later brought to life Willis O'Brien's *King Kong* (1933) and Ray Harryhausen's giant bee in *Mysterious Island* (1961) and scorpions in *Clash of the Titans* (1981). The process of filming an object frame by frame so that it appears animated when the frames are played back dates

Akihiro Higuchi uses insects as his canvas, and his *Mitate-Urushi* (*Ko121*) is a pair of decorated stag beetles.

to the final years of the 1800s, but clever experiments took a narrative leap soon after this with insect-fueled works by Polish-Russian animator, Ladislas Starewitch.

Starewitch (also Starewicz, Starevitch, or Starevich) began by animating a pair of stag beetles, using wax to stick wire legs on actual specimens. The animation, since lost, took the form of a stag beetle battle, resulting in the earliest stop-motion puppetry in film history. This one-minute production was Starewitch's way of fabricating a behavior he could not film with live beetles because they refused to perform under bright camera lights. Plots thickened in his many animations that followed. Swords are swung, canons fired, axes brandished, and castle explosives detonated, all by beetles and all over the love of a

female in *The Beautiful Leukanida* (1912). In *The Cameraman's Revenge* (1912), infidelity runs rampant as Mr. Beetle and Mrs. Beetle independently have affairs, but one ends with a portrait of Mrs. Beetle crashing over her lover's head and the other with a screening of Mr. Beetle's own indiscretions in a movie theater. An homage to Aesop's fable, *The Grasshopper and the Ant* (1912), ends in a lonely death for a carefree grasshopper facing the impending cold of winter after his entreaties to industrious ants were rebuffed. What started as a creative solution to filming a behavior too difficult with live beetles turned into an influential career of stop-motion animation, with some of his greatest stars being insects. So new was this technique, and so ignorant his audience to insect behavior, that some thought he had trained his insects to perform like humans.

Training insects for art is not unheard of. María Fernanda Cardoso captured the theatricality of the flea circuses of yesteryear by training her fleas for performance art pieces. Professor Cardoso and The Queen of the Fleas (her stage name) designed miniature props and arenas within which

her female cat fleas performed as "Flea Jugglers, Brutus the Strongest Flea on Earth, Tightrope Artists, the Flea Cannonballs, Alfredo the High Dive Artist, Trapeze Artists Sarindar and Dimitri, as well as the Flea Ballerinas dancing on a musical box." Cardoso had never worked with fleas, and had no one to train her, so she learned the ways of the fleas by trial and error, and fed them . . . offering her hand and her blood as sustenance.

Miniature props and trained fleas are all part of *Cardoso Flea Circus*, with Professor María Fernanda Cardoso as Queen of the Fleas at The Fabric Workshop and Museum, Philadelphia.

Working with living insects poses challenges, but artists like Cardoso and Yanagi, who know their insects well, can reap creative rewards. The same is true for Kazuo Kadonaga, Xu Bing, and Tera Galanti and their silkworm moths, Aki Inomata and her bagworm moths, and Hubert Duprat and his caddisflies. The same is also true for two artists independently working with leaf-cutter ants, those abundant rainforest farmers who carry cut foliage like parasols down from the canopy to their fungus gardens.

Donna Conlon enticed a colony of leaf-cutter ants in Panama to carry tiny flags and symbols of peace (*Coexistence*, 2003), while Catherine Chalmers, during a decade of recording and experimenting with colonies in Costa Rica, followed two colonies into combat. As the ants battled, Chalmers documented the protracted conflict against a stark white background in *War* (2021). She gave me hints about her approach and method: "I watch, respond, experiment, and adjust. Through this back-and-forth process, the project took shape. Patience is a primary color on my palette; force and manipulation are not." Observe insects carefully, as Chalmers has, and you can discern personalities across individuals, or across colonies: "I observed a striking variation amongst ant colonies in their response to the presence of another colony harvesting in the vicinity. Some were tolerant of their neighbors and others were quick to wage war. It seems that even with ants, some groups are peaceful and others are aggressive."

Catherine Chalmers's scenes of tragedy, one staged without fatalities, *Executions: Gas Chamber* (2008), and the other consequences of battle in a Costa Rican rainforest, *War* (2021). Chalmers filmed American cockroaches (*Periplaneta americana*) as if they were suffering human forms of execution and filmed two colonies of leaf-cutter ants (*Atta cephalotes*) waging war over days. Chalmers set white plastic sheets on the forest floor in preparation for the continuing drama, dismemberment, and death.

As with all insect media art, sometimes the results are whimsical, and sometimes they are profoundly inspiring, or unsettling. Insect fairies can be all of the above, as we see in William Shakespeare's *Romeo and Juliet* (1597). Mercutio describes a carriage of the fairy queen, driven by a fly and composed, in part, of spiderwebs, grasshopper wings, and cricket bones.

Her waggon spokes are made of spinners webs,
The couer, of the winges of Grasshoppers,
The traces are the Moone-shine watrie beames,
The collers crickets bones, the lash of filmes,
Her waggoner is a small gray coated flie,

Picture walking into a room in which tiny insects seem to be hovering in the corner. Approach more closely and you might be struck by what Catriona McAra calls "an entirely different register of reality." Minute skeletons bearing wings dangle in midair. These tiny fairies wield weapons consisting of sea urchin spines, hedgehog spines, an insect foot, wasp stingers, or an earwig. In another scene, they hurl ants from a flying cicada, grasp claws as grappling hooks, or use a jewel wasp as a drill. The fairies resemble miniature human skeletons, but with wings. They ride bumblebees, taxidermy mammals, or travel in skulls, even into outer space.

These are the worlds of Tessa Farmer, British artist and mythmaker who constructs each fairy ("too real to be untrue," according to Giovanni Aloi) from a plant root (bird's nest fern, *Asplenium nidus*), securing it with superglue, and insect wings. The wings are mostly from flies (Diptera) or wasps and their close relatives (Hymenoptera).

Different writers grapple with understanding the unique installations by Farmer in the book *In Fairyland: The World of Tessa Farmer*. "They could be the dead, fallen angels, the relics of an ancient race, forgotten gods, or simply another dimension of beings that interacts with our own," proposes Gail-Nina Anderson. "They are anarchic and in constant revolt, similar to cells that separated themselves from the unifying organism. They are the organs without body, the deserting angels," or "the missing link between

Five "fairies" ride atop a damselfly in *The Perilous Pursuit of a Python* (2013), while another wears ant clothing when approaching ants in *The Coming of the Fairies* (2011). These are small components within larger installations of taxidermy, bones, insects, plant roots, worm shells, and more. Tessa Farmer constructs these "fairies" with plant roots and insect wings, slightly more easily seen in the image of the lone fairy perched on the back of a bumblebee.

the human and the insect worlds," speculates Petra Lange-Berndt. Gavin Broad, museum curator and entomologist at the Natural History Museum in London, who has carefully examined Farmer's fairy worlds for their insect origins, finds elements of evolutionary truths. Some of the horrors that lurk in ecological interactions all around us seem present here, but with a disturbing twist. "This corrupted, skeletal humanity imposed on insect dimensions, and often part-insect physiognomy, imparts a willfully malicious direction to their actions." I am privileged to own one of Farmer's fairies, and each time I look at it standing atop a hovering bumblebee, I can't help but be transported to another reality, one that mirrors the strange symbiotic relationships, documented or unknown to us, hidden in our midst.

How often do insects and their products appear in art, and what insects are most often represented? Countless works are colored with cochineal, coated in shellac, sewn with silk, and infused with beeswax. Among these, silk may be the most common insect-derived medium in the world of art, though no one has attempted to census the proportion or absolute number of art pieces created or coated with any of these common ingredients. And it would be silly to consider the countless Van Gogh grasshopper accidents in art's history, when the intent to add insect media was neither the artist's nor the insects'. I decided to survey a tiny subset of insect media art in which artists explicitly and intentionally used insect media, and discovered that among 164 works, 65 were made from insect products, and the remainder were evenly split (33 examples of each) between art made using insect body parts, dead insects, or living insects. Fifteen different orders of insects populated the art, and although beetles are the most diverse order of insects and we probably interact most with flies, I found the order including bees, wasps, and ants (Hymenoptera) to be the most popular subjects, followed by the order of moths and butterflies (Lepidoptera). This artistic bias was driven by the long and rich history we have enjoyed with two species: the western honey bee (*Apis mellifera*) and domesticated silkworm moth (*Bombyx mori*).

Human culture is sewn, colored, and fed by the innards of insects. We work with them, eat them, fashion traditional medicines with them, and create art using any and all elements of them. They will undoubtedly travel with us in our future space voyages to inhabit other terrains and further serve our needs. Our dependence on insects runs deep, though we don't always realize it. As we saw with people attempting to replace insect pollinators with chicken feather dusters, attempts to fashion surrogates for insects can expose our need for them. If those serving vital ecosystem functions disappear, could we replace them? Could we mimic their components with technology, or fabricate competent insect robots to find missing people or serve as spies? What reasons might artists or engineers have for creating insect facsimiles? When working with real insects—the topic of our first section—is untenable or undesirable, humans have chosen to make them.

MAKING THEM: GENESIS

After months of intense work, the team of entomologists assembled at the Pentagon in Q3 to unveil their Mach 1 prototype. Senior Entomologist Dr. Gilford Clavis led the assembly, walked around the room, making sweeping gestures as he drew them in closer: "What you are about to see is the greatest synthesis of hundreds of millions of years of insect evolution with state-of-the-art human engineering. A harbinger of a robot army that no one will be prepared for. Please welcome Rob O'Fly." From behind him, emerging from his shadow, came Rob, a thirteen-year-old teenage boy. "Hey, dudes," uttered the teen, voice cracking. He took one awkward step toward the assembled group, shoulders hunched forward, hair hiding the battlefield of pockmarked pimples on his forehead. They all gasped, shrinking back. One colonel fainted and had to be carried away.

Dr. Clavis reassured them that what they were witnessing was, indeed, a teenage boy, albeit with significant enhancements. "This unassuming boy before you has the blood of a fire ant coursing through his arteries, has been stung in the butt by a giant hornet, but has buttocks reinforced with the elytra of a titan beetle, has the problem-solving ability of a tarantula hawk wasp, the acerbic wit of a scorpion fly, the smug equanimity of an assassin bug, and the courage and generosity of a praying mantis. And let me not forget, he also comes equipped with the digestive tract of a male butterfly, the genitals of a honey bee drone, and the lifespan of a mayfly!"

Upon hearing this, the gawky teen dropped in a heap to the floor.

—ARNO KLEIN, 19 FEBRUARY 2023

It's tempting to think of evolution as an engineer, and the finely honed features of organisms as the products of either a mastermind machinist or the result of some well laid-out plan. Instead, evolution acts more like a tinkerer, using what is available to produce traits without a preconceived path. If successful, the traits promote survival, but they are never perfect. If perfection were the norm, it would be odd to live among only 1 percent of all species to have ever evolved on Earth, with the remaining 99 percent of what some estimate to be four billion species extinct.

Perfection is unattainable because, at least for life, it is meaningless. With every trait comes trade-offs, and what is a boon in one context can be a bane in another. Ultraviolet light from the sun helps us synthesize vitamin D, which strengthens our bones, but those same rays damage our skin. How much sunlight should we get? That depends on how far from the equator you are, what season it is, and how much protective melanin is in your skin. The darker your skin, the greater your protection, but the greater your need for synthesizing vitamin D. Now consider your huge brain, and the trade-off this presented when your biological mother gave birth to you. Developing with big brains means painful, dangerous births. Male long-tailed widowbirds (*Euplectes progne*) attract mates with their extraordinarily long tail feathers. The longer a male's tail feathers are, the more mates he will attract, but the harder it will be for him to escape predators. If you were to glue on feather extensions, as did biologist Malte Andersson, females would find the preternaturally long feathers to be far more seductive than anything they could ever find in the wild, but unwieldy tails are costly burdens. If every physical aspect of every living being is a manifestation of trade-offs, it is easy to imagine a dynamic game of thrones, where plants compete for light, a parasite coevolves with host, and victory depends on a delicate balance of variables. Seats of power and influence are always in flux in a changing environment.

Insects face trade-offs all the time. For an insect to breathe, spiracles on the sides of the body must open to let in oxygen-rich air and expel carbon dioxide, but to keep them open means losing precious water. Their exoskeletons must be strong yet flexible. One of their biggest trade-offs comes with being small.

Being big has its advantages, but try laying an egg inside the egg of another insect. There are many nutritious eggs out there, but to take advantage of insect eggs as hosts for your young is the province of the minutest of the minute. Getting that wee, however, means the insects have to make sacrifices. Their exoskeletons are thinner, some muscles are missing, and circulatory and respiratory systems are greatly simplified. Some insects lack a heart, and the brain of a corylophid beetle is squished inside its thorax. The smallest insects are the size of single-celled organisms, and are among the smallest animals in existence. By body length and volume, only a select number of roundworms, polychaete worms, worm-like gastrotrichs, and rotifers can get more diminutive. The smallest insect ever described is a male "fairyfly," the wasp *Dicopomorpha echmepterygis*. Wingless, eyeless, lacking mouthparts, and measuring as little as 139 micrometers long, three can fit inside the egg of a bark louse within which a female fairyfly of this species lays her eggs. This lumbering, disproportionately long-legged male is a smidge longer than the thickness of a sheet of paper or the width of a strand of human hair, both measuring about 100 micrometers. The larger female fairyfly is still so small that flying through the air to locate her hosts' eggs is more like swimming, with aerodynamics relying on different physics entirely than experienced by her far larger wasp relatives.

The very smallest flying animals on Earth belong to *Megaphragma*, another group of parasitoid wasps, and Russian entomologist Alexey Polilov has discovered that *Megaphragma mymaripenne* has evolved a most unusual way to downsize. When transitioning from pupa to adult, the wasp loses almost all (greater than 95 percent) of her nerve cells' nuclei, more than 7,000 of the 7,400 she had as a pupa. Why should this matter? The nucleus is the control center of an animal's cell, and is required for cell reproduction, and manufacturing proteins. We depend on nuclei in almost all of our body's

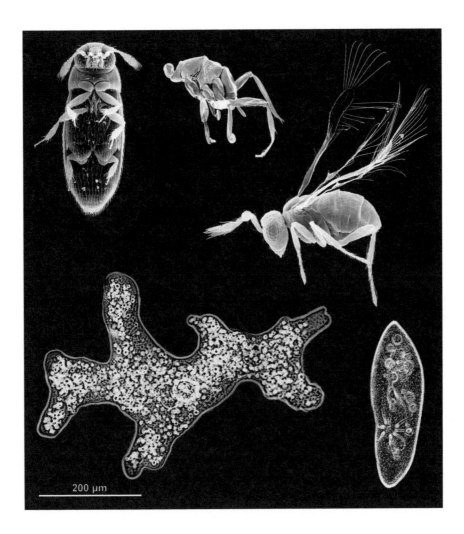

200 μm

The smallest known insects, alongside similarly sized single-celled organisms! *Nanosella* sp., the beetle (*upper left*), is the smallest non-parasitic insect. *Megaphragma mymaripenne*, the winged "fairyfly" (*right*), is the smallest flying insect. *Dicopomorpha echmepterygis*, the wingless male of an egg parasitoid (*top center*) is the smallest known insect on the planet. Alexey Polilov studies these minutest of insects, and cleverly concocted this visual collage of insects beside organisms that live life as a single cell, *Amoeba proteus* (*bottom left*) and *Paramecium caudatum* (*right*).

cells, and when we lack them, it had better be for a good reason. Our red blood cells lack nuclei to make room for oxygen-toting hemoglobin, and when they die our bone marrow can quickly replace them. Almost none of *Megaphragma's* neurons can generate more of themselves. This probably contributes to her expiring after only five days, though by lacking nuclei, she can live the specialized life of the Lilliputian. Her brain volume is almost halved when she transforms into an adult, and these wasps have the fewest number of neurons of any insect (4,600 in her adult brain, versus 850,000 in a honey bee's brain and 86 billion in our brains). Despite extremes of miniaturization, micro-insects invest up to 16 percent of their body mass to brains (we invest only about 2 percent). Brains are costly and, as expressed by scientists Joseph Kilmer and Rafael Rodríguez when studying miniature spiders, "small animals pay relatively more for absolutely less."

We know almost nothing about their behavior (try observing something so little), but we can say that micro-insects manage to eat, to fly, and, in the case of *Dicopomorpha* and *Megaphragma* females, to find the eggs of their hosts. Though it may not look to us like these minuscule, short-lived insects lead rich lives, they have successfully filled niches and survive on a scale beyond our awareness. In doing so, these animals are tiny triumphs, and can offer us insights into the limits of miniaturization.

Whether small or large, the hundred myriad known species of insects have weathered the uncompromising forces of selection by nature for hundreds of millions of years. As a consequence, insects' colors span the rainbow, some can sense a forest fire miles away or navigate by polarized light, and limbs are superbly specialized for moving in wind, water, or soil. They thrive socially or in isolation, and have adapted to extreme living conditions. Each insect offers lessons waiting to be learned that could affect how we conduct our lives. In our first section, we saw how insects' bodies and bodily products have helped shape our world, but it isn't always convenient or desirable to exploit insects directly. Here, we will explore examples of when it behooves us to build them instead. The artist and the engineer both borrow from insects' body plans, and follow an intuition that an insects' construction can instruct us.

Exoskeleton Envy

My son Rivyn likes to make cardboard helmets to complement his body armor. The material is relatively stiff and unforgiving, so as his body grows, he will need to shed the little pieces of armor that have served him so well in battles against imaginary beasts. Should the creatures continue their epic onslaught, he may need to forge a new protective suit to ensure the safety of the lands he has populated with handmade puppet dinosaurs, bees, and boggarts. When I think of Rivyn accommodating his growing body in this way, I see a human adopting an insect's solution to becoming larger. Insect armor is made primarily of cross-linked proteins with chitin—simple sugars bound together in chains—and must be shed after a new, larger exoskeleton is generated below. After shedding the old exoskeleton, the new one expands to replace it, and hardens with time. It is a perilous transition for the insect, but a necessary one to allow its armored body to grow. Although an insect's process of generating armor is different than ours, an insect is sheathed in protective plates, much like the plate harnesses worn by knights centuries ago in royal courts and on blood-soaked fields of battle.

No place is our armor more obviously inspired by insects than in the horned beetles, voracious mantids, and aerobatic dragonflies decorating Japanese helmets (*kabuto*). Early samurai helmets display subtle insect imagery, but after guns infiltrated war efforts and clouded battlefields with smoke, the insects became more elaborate, and more visible, on front crests (*maetate*). During the Edo period, a time of peace, *kabuto* escalated in their extravagance, and became showpieces more than battle-hardy military equipment.

Adorned with insects or not, human body armor exudes insect envy. To cover our relatively soft bodies with an impenetrable exoskeleton is to become superhuman . . . save for the fact that armor can be heavy, clunky, and can reduce vision. Wearing a helmet, visor, or shield can protect one's eyes, but at the expense of constraining visual range and clarity. Dung beetles with head armaments face an analogous trade-off, in that they can either develop with large horns and small eyes, or small horns and large eyes. If we were to truly mimic an insect's exoskeleton, our armor would serve functions well beyond protection, being full of sensory apparatus and versatile in its flexibility. The exoskeleton has been one of the great evolutionary innovations contributing to insects' success, and technicians, architects, and engineers have taken notice.

One source of exoskeletal inspiration can be found leaving tiny footprints on the dunes of the Namib. Picture yourself there:

You are in the oldest, driest desert in the world, surrounded by stretches of sand dunes—and nothing more. With time ticking and your parched body requiring the one molecule like no other that could sustain it, you scan for sources of water. How can anything survive in this vast, barren place? Dawn breaks and a morning fog drifts through your outstretched fingers. Water! Harvesting this precious vapor seems impossible, but look to your feet and beetles stand with their posteriors in the air, basking in the fog. Water condenses on bumps protruding from the darkling beetles' slick, black exoskeletons, and slips down along channels into their open, waiting mouthparts. Like water off a duck's back, the water forms beads on the waxy surface of the beetles' wing covers until each bead reaches its maximum size and slides down, making room for a new droplet to form. Adopting this water-harvesting ability into a scaled-up, sustainable, energy-free strategy has been on the minds of engineers designing drinking fountains, self-filling water bottles, underground watering systems, water-harvesting mesh nets, synthetic surfaces, and entire buildings. With names like Warka Water and FogQuest, projects to passively collect water from the air inspired by fog-basking beetles hold promise for communities lacking viable sources of drinking water. When adversity strikes, it often helps to observe and study those who have already worked out the solutions. Here, a small bumpy beetle that stands on his head can teach us how to passively extract water from the desert air.

Decorative samurai helmets featuring insects, all by anonymous Japanese artists. The dragonfly, the favorite emblem of the samurai, appears with vertical abdomen in *Helmet in Dragonfly Shape* from the seventeenth century (*above, left*; made of iron, lacquer, wood, leather, gilt, pigments, silk, papier-mâché), and flying above clouds in *Helmet (Kawari-kabuto) Surmounted by a Dragonfly* from the eighteenth century (*above, right*; restored 2015, and made of iron, copper, gold, silver, wood, lacquer, silk, linen, hemp). The cicada helmet (*right*), with gilded eyes, is made of iron, gold, papier-mâché, lacquer, silk, and textile (1615–1868).

Inspiration to advance technologies has frequently come from beetles. The diabolical ironclad beetle, which can survive absurdly strong crushing forces—being run over by a car is a minor inconvenience—is the stuff of dreams for an engineer wishing to create indestructible materials. A weevil from the Amazon that shimmers with a green iridescence from nearly any viewing angle can lead a prepared mind to think of photonic crystals and optical computing. With but the touch of a finger, a gold tortoise beetle changes color, a potential attribute for printable, sensor-laden robot skins. Bombardier beetles, noted for their ability to explosively discharge boiling quinones from their rear end when provoked, have inspired scientists to test applications for gas turbine igniters.

All of these beetle-derived technologies are examples of biomimicry—basing our inventions on some aspect of a living being. Think of every human who has ever watched a bird or a bat and wished they, too, could flap, hover, and glide. Four hundred years before Orville and Wilbur Wright gave humans the power to fly by designing and building a motor-operated airplane (1903), Leonardo da Vinci drew plans for flying machines based on the wings of birds and bats, and insects. One of da Vinci's most complicated designs imitates the movement of dragonfly wings. As we will see shortly, some of today's cutting-edge micro-aerial robots and drones are based on the biology of insects.

Even our most iconic and revolutionary invention of all time, the wheel, may have ties to insects. Though the wheel's origins are unknown, early evidence for potter's wheels dates to at least 4000 BCE in Mesopotamia. Someone somewhere first had the insight to spin a wheel on an axle to transport heavy weights or to form symmetrical ceramics. Did that person experience an epiphany utterly out of the blue, or was that person an astute observer of the humble dung beetle? According to biologist Gerhard Scholtz, the invention of the wheel in human culture "was merely a reinvention, copied from nature and from dung beetles in particular." Before you scoff at the notion that humans simply reinvented one of our culture's most pivotal creations, consider Scholtz's rationale: (1) Domestication of animals led to greater densities of dung beetles attracted to those animals' dung, making the beetles

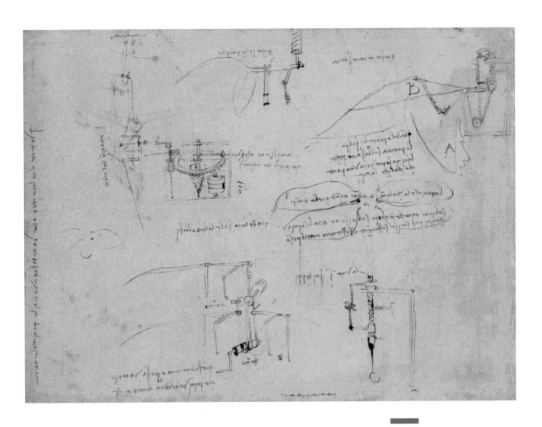

accessible for observation; (2) having domesticated animals with the ability to pull heavy weights made it easier to see why something like wheeled plows and wagons would be helpful. Add to this the fact that dung ball–rolling beetles are found on almost all continents, and that few other organisms engage in wheel-rolling motions.

Leonardo da Vinci modeled a flying machine after a dragonfly (1485).

Wheels are old news, and the role of the dung beetle in its invention is mere speculation. Let's look at more recent engineering advances that give a clearer view of the impact of insects across technologies. An engineer prospecting for influential features of an insect need only travel from one end of the insect's body to the other to uncover treasures worth emulating. An adult insect's body comes in three major segments: head, thorax, and abdomen, though a bit of the abdomen has migrated to the middle segment in

Hymenoptera (bees, wasps, ants, and sawflies), and two-thirds of a beetle's thorax has moved in the opposite direction, and is conjoined with the abdomen. Traveling the landscape of an insect's head, thorax, and abdomen, we find striking anatomical attributes that have profoundly shaped our history and material culture. Sometimes, artistic license has corrupted the insect into forms deemed more appealing (children's cartoons) or more frightening (adult horror) to human audiences. Engineers, however, are not looking for abstractions or exaggerations, but real, potentially influential features. By seeing insects for what they are—fellow animals trying to survive, and evolving marvelous ways of doing so that have a net positive effect on our own existence—we stand a chance of appreciating nature, and learning how to copy it.

Head

An insect faces the world head-on. Sensing what is ahead with a head makes sense when it is equipped with organs capable of sight, smell, taste, touch, and, sometimes, hearing. Though we will learn how insect mouthparts could teach us to design better medical devices later, what attracts the most biomimicry attention lies in insects' intricate compound eyes. Insect eyes may not produce the resolution our human eyes can, but they do succeed in ways ours do not, and on a much smaller scale. Try to sneak up on a fly to be reminded of why there are so many buzzing about. They are far better than we are at detecting motion across their many lenses. Insect eyes also have a wide-angle field of view and infinite depth of field, producing minimal aberration. Nicolas Franceschini and colleagues at the French National Center for Scientific Research (CNRS) took an early interest in the fly's compound eyes in 1992, and attempted to share insights about these "masterpieces of integrated optics and neural design" by creating a robot bearing the artificial eye of a fly.

There are many ways to build an eye, and the future of drones and medical probes may rely on our getting as close as possible to what has proved so successful in nature. Fourteen years after Franceschini's fly-eyed robot, engineers at UC Berkeley, Ki-Hun Jeong, Jaeyoun Kim, and Luke Lee, took the

challenge of imitating a compound eye to another level, building individual lenses out of photosensitive polymer resin on the scale of tens of micrometers, half the width of a typical human hair. Working on a much larger scale, Young Min Song and fellow engineers collaborated at the University of Illinois Urbana-Champaign to create a digital camera composed of 180 elastomeric elements on a curvilinear space backed by silicon photodetectors. The camera, with all of its insect-inspired lenses, can record across 160 degrees, and nothing goes out of focus. Though the resolution is low, future versions could exceed that processed by actual insect eyes. Song is currently developing cameras based on the eyes of crabs, cuttlefish, and fish.

Digital camera that takes the form of a hemispherical compound eye (circuit board and external electronics not shown).

Figuring out how a compound eye works often starts with experimentally testing an insect's behavior. When a worker honey bee flies back from a food site to her nest and advertises the site to her nestmates, she communicates the distance they will need to travel by waggling her posterior for a specific duration, turns around, and repeats the waggle for the same duration, again and again. The duration spent waggling depends on optic flow—the amount of visual detail that crossed her eyes during the flight. When moving through the world, the bee experiences apparent motion of visual features, like the edges of objects getting larger as she gets closer. More visual complexity is perceived by the brain as a greater distance traveled. It doesn't matter how much energy she expends during the flight; it is what flows across her eye that she uses when communicating distance.

Neurobiologist and engineer Mandyam "Srini" Srinivasan, fellow Australian biologist Shaowu Zhang, and German colleagues determined this by designing an ingenious series of experiments. They trained honey bees to fly from their hives to food they provided, but sometimes the bees had to fly an additional twenty feet (6 m) through a short and narrow tunnel to reach the food. The top of the tunnel allowed sunlight through and, when the tunnel was decorated inside with random black-and-white patterns, the bees perceived the distance traveled as farther than when the tunnel was decorated with stripes running parallel to the length of the tunnel, which shouldn't affect optic flow. Srini and the others knew this was the case because the bees returning to their hives performed dances indicating greater distances from hive to the food site. When trained to feeders at different distances from their hives that included a final flight through the decorated tunnel, bees massively overestimated their flight distance when communicating the food sites to nestmates. Each phase of their dance (when they waggle their bodies back and forth as they move along the honeycomb) lasted longer than it should have. Traveling less than twenty feet (6 m) in the decorated tunnel was perceived by the foraging bee as equivalent to an average of flying 610 feet (186 m) outdoors.

Optic flow is what mattered, and engineers have concocted different ways that artificial eyes can detect optic flow. Dario Floreano, director of the Laboratory of Intelligent Systems at the Swiss Federal Institute of Technology Lausanne has been involved in several of these efforts, having developed a compound-eye-inspired hemispherical array of microlenses, and a system of multiple "elementary eyes," each two-thousandths of a gram and capable of detecting motion from multiple directions. Eye-imitative technologies are developing quickly, and capitalizing on eyes that have survived hundreds of millions of years will enhance our machines of medicine, of peace, of surveillance, of war.

Thorax

Nestled right behind the head, *Ormia ochracea* flies have ears. It took much effort for biologist Daniel Robert to find them when he was working in neurobiologist Ron Hoy's lab at Cornell University. No fly had ever been shown to have an ear that could detect sound from far away, but Robert knew *Ormia* must, because females fly to recordings of cricket calls when played outside. No stimulus other than the stridulating songs of males lured them to the speakers. The fly's purpose for flying to the songs of another species is simple: Lay larvae on cricket hosts, or your own larvae will eat you from within. But the mother fly needs to locate the singing male cricket faster than a female cricket does, otherwise the male will stop singing and start mating. To find singing male crickets from far away, the flies use strange, and strangely efficient, ears at the front of their thorax. The ears should be too small and too close together to work so well to locate distant sound sources, but the ears have a flexible bridge connecting them and as the bridge bends or rocks, it exaggerates the time taken for sound to reach one ear versus the other ear, and the loudness perceived by each ear. If the sound arrives sooner and is louder in the left ear, that ear is closer to whatever is producing the sound. The fly is able to navigate to the song with the accuracy matching that of a human with healthy hearing. As we age and our hearing fades, we lose this ability to accurately orient toward sound sources, so these remarkable fly ears are the source of inspiration for the design of miniature directional microphones and hearing aids.

The thorax can have ears, but this is the body region most famous for bearing legs, wings, and the muscles to power them. Just as our arms and legs allow us to embrace loved ones, escape danger, and perform tasks both delicate and forceful, insects' appendages make it possible for them to grasp, flee, or work. The power, dexterity, and speed of insects, coordinated and executed by these appendages, are largely what leaves the greatest impression on us. One pair of segmented, jointed legs connect to each of three thoracic segments, and Earth's oldest sets of wings arise from the second and third of these segments. Beyond this generality, insect legs and wings have specialized in outrageously variable ways. The patterns and hues of a butterfly's wings, the versatility of a dragonfly's flight, the grand leap of a flea, and the gait of a cockroach have all inspired human technology. We will continue our tour of the thorax with legs, and sticky lessons from a pair of leaf beetles.

Legs walk, run, swim, hop, grab, clasp, and burrow. Sometimes, they stick.

Holding on can be difficult, especially if you find yourself upside-down, walking vertically or on a slippery surface, or when someone is trying to pry you from your perch. Picture yourself as a palmetto tortoise beetle, *Hemisphaerota cyanea*. You began your life as an egg, generously protected with fecal pellets deposited in a hardened matrix by your mother. Minutes after you hatched as a larva, you began eating a palmetto leaf, and shortly after that, you experienced the inevitable urge to defecate. Instead of voiding your excreta and moving on, you do what other tortoise beetle larvae do—you hold on to it, actually gluing it to a forked projection on your abdomen. In your case, however, the strands keep coming and coiling above you, forming a spectacular nest-like wig through which your body can no longer be seen. Each time you molt, you keep the strands. When you pupate, you pupate beneath them.

When you emerge as an adult, you expose yourself only after your exoskeleton safely hardens. Your wild wig-shield protects you from predatory ants, stinkbugs, and ladybird beetle larvae. Though a species of ground beetle can overcome this defense, you were spared from their crushing mandibles, and amble about these days in the adorable manner of beetles resembling turtles. The fecal shield is but a memory. Now that you have matured beyond the indecorous behavior of your youth (mating and flying would be difficult with the extraneous excrement), you need a new way to defend yourself, especially against omnipresent ants. Ants could pluck you off a palmetto, hoist you up, and carry you home for dinner, except that you have a tarsal trick up your sleeve.

On each of your six feet (tarsi), you have ten thousand bristles, and each bristle ends in a forked pair of pads. If disturbed, your first impulse is to hunker down, pressing all sixty thousand bristles against the leaf on which you stand. The flattened brushes of your feet increase the contact area between feet and leaf, but to make your feet stick, you produce oil through microscopic pores between the bases of the bristles. Like a drop of water between two plates of glass, the film of oil makes pulling your body off the palmetto a major challenge for any hungry ant. On a good day, you might be able to withstand a tug of two grams, the equivalent of 148 times your body mass. Thomas Eisner, the clever and curious entomologist who studied this remarkable behavior, compared this to a 155-pound (70.3-kg) human pulling 23,000 pounds (10.4 mt), or seven and a half Subaru Legacy station wagons. Eisner found the record (by attaching weights with pulleys) to be 3.2 grams, or nearly 240 times the beetle's body mass. This display of power lasts only seconds, but because ants give up tugging after about 23 seconds, the fact that you can hold on with a pull of 0.8 grams (nearly 60 times your body mass) for 2 minutes means that your relatives rarely end up on an ant's menu. After the ants give up, you peel your feet up and amble on.

I'm guessing you know where this is heading. Engineers saw the value in fabricating a switchable adhesive device, so adopted two lessons from these bristly, oily beetle feet: produce surface tension with many small liquid bridges, and make the stickiness quickly reversible (switching on or off by electronic control). They published their proof of concept imitation tortoise beetle feet in 2010, one year before Eisner died.

Another beetle in the same family (Chrysomelidae) can also stick to surfaces with oil-covered hairy pads for feet, but *Gastrophysa viridula* ups the ante by sticking to surfaces underwater. Oily, hairy feet normally stick well only to dry surfaces, so the beetles need to do something beyond what the tortoise beetle does to be able to stick to slick leaves or stones when submerged. The secret is in bubbles. Their feet trap air bubbles, which serve to de-wet the surfaces on which they walk. The air bubbles are held between the bristles as they temporarily dry surfaces where beetles step, and work with the oily bristles to create adhesion. Discovery of this led to our second artificial beetle foot design. Naoe Hosoda from the National Institute for Materials Science in Japan and Stanislav Gorb at the University of Kiel may be the first biomimetics professionals to publish a paper featuring a toy bulldozer underwater. Covered by an artificial silicone polymer, the bulldozer's wheels trapped air while stuck on inclines and vertical surfaces, demonstrating the same principle as championed by their leaf beetle model.

Bulldozer beetle wheel feet may seem a surprising way to honor insect legs, but the thorax's other appendages offer comparably bizarre biomimetic surprises.

Flapping and fluttering, wings are the most iconic asset of the adult insect. Only two orders of insects have never had wings in their history: Zygentoma (silverfish and firebrats) and Archaeognatha (jumping bristletails), though some other insect lineages have evolutionarily forsaken wings over time. There is a reason that going wingless is an exception for adult insects. Wings are not only the tools for flight, but they can also provide camouflage, give warning, attract mates, help with thermoregulation, repel water and microbes, or serve as musical instruments. Each set of wings is unique, and represents a triumph of surviving nature's incessant and pitiless pressures. Identifying the reasons for each wing's success keeps entrepreneurs with a bent for biomimicry busy.

Aside from the obvious allure of fabricating insect wings for flight (see robots, coming soon), humans are drawn to the extravagant colors of butterfly wings, and the brilliant blues of *Morpho* butterflies in particular. Finding blue pigments in nature is rare. Lapis lazuli, mined in Afghanistan to produce ultramarine, is one of the only examples. Yet *Morpho* butterflies shimmer with a range of exquisitely iridescent blues throughout the neotropics. Only recently have scientists figured out how these blues are produced.

Marco Antonio Giraldo, a biophysicist in Colombia, collaborated with colleagues to unveil two layers of wing scales that, together, reflect different blues in a way that largely matches the evolutionary groupings within the lineage of *Morpho* butterflies. The scales on the surface are stacked with slender, textured plates. The spacing between the layers determines the hue, and the number of layers within each scale determines the intensity. A layer

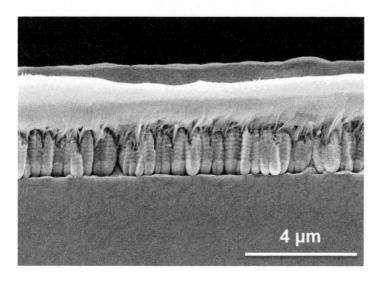

of brown scales lies below and serves as thin-film reflectors. Though iridescence typically results in colors that change at different viewing angles, *Morpho* iridescence delivers a stable, bright blue across a wide range of viewing angles. Duplicating this intricate color-producing paradox has been challenging, but techniques adopting a "hierarchy of disorder" found in their scales appear to be homing in on the once elusively stable iridescent blue. The structures causing insect iridescence have already inspired shimmering fabrics and fashion, and e-readers, which save battery power by reflecting light rather than transmitting it.

The erratic blue flurry that is a *Morpho* in flight appears without warning. The butterfly bobs, glides, and flutters, then disappears. This magical quality to startle, then vanish, is possible because the underside of each wing is dull and camouflages beautifully in the forest when the wings fold and the butterfly is at rest. To appear dull means to reduce the amount of bright light you reflect, and plenty of insects accomplish this with antiglare structures on their exoskeletons. Glare is when vision is obscured by bright light, and it can be reduced if surfaces are less reflective. Eyeglasses and windshields are easier to see through when coated with an antiglare film, and antiglare surfaces on computer screens increase contrast and clarity to enhance visibility. Photovoltaic cells in solar panels absorb more light energy when less of the incoming light is reflected away. For each of these technologies, engineers are emulating insects and their ability to remain incognito. The surfaces of cicada wings and clearwing butterfly wings, as well as of moth eyes and fruit fly eyes, all bear antireflective features that likely make them less conspicuous to predators, but more attractive to engineers.

Another way to become less conspicuous is to become smaller. Some insects roll up, bend in their legs, or fold up their wings. Mayflies (Ephemeroptera) and dragonflies and damselflies (Odonata) do not share this feature, so their wings jut out, sometimes inconveniently. The moment insects evolved the ability to flex their wings over their abdomens (thanks to a tiny muscle pulling a tiny armored plate at the base of the wings), winged insects had the option to crawl under bark, scurry under rocks, slip into burrows, or swim through water. Folding wings, like origami, the Japanese art of

folding paper, can push the limits of compact storage, and inform technologies that could benefit from folding. So appropriate is the analogy of wing folding with origami that a scientific report showing the intricate folding of ladybird beetle's wings visualized the process by comparing the actual wing folding with a sequence of illustrations showing folding in the style of origami.

Flying Kabutomushi (second version), designed and folded from one 22-in. (55.8-cm) square of tracing paper by Shuki Kato in 2011 (wingspan: 9.5 in. / 24.1 cm). Kato is featured in a book of origami masters about the Bug Wars.

Origami involves transforming a material using folds, without ever adding to its volume or weight. Though an ancient art, origami is a dynamic, contemporary practice, and informal competitions (Bug Wars) have birthed some of the most complex and realistic insect origami ever created. The real-life world champion of folding is the earwig (Dermaptera). Earwigs store membranous hindwings under a pair of tiny, hard front wings, until ready to fly. The front wings lift up and the hindwings unfold into ten times their original size.

The wings lock in place, stabilize without the need for muscle power, and are ready for flight. No other animal can do this. The elegant unveiling of wings and flight have been captured at six thousand frames per second by entomologist Adrian Smith, and engineers from ETH Zürich and Purdue University have 3D-printed prototypes following the principles of earwig wing folding. Though no one has yet designed a self-folding structure as complex as the earwig wing, some of these simpler prototypes hint at our future of folding tents, maps, umbrellas, packaging, satellites, and solar sails. The elegant art of origami finds applications even in our bodies, with designs for unfolding heart stents, and the potential to produce other self-expanding medical devices.

When designing medical instruments, it can help to look in unlikely places for inspiration, and insect wings, once again, can motivate us to rethink our technology. Developing a medical device for a large primate (that's what we are, after all) using a tiny, spineless arthropod as a model seems strange at first. That is, until you look—sometimes through a microscope—at what an insect can do, or how it's built. The tiniest of protrusions on the wings of dragonflies and cicadas, for example, could present a solution to the growing threat of antibiotic resistance.

The problem: Many bacteria pose health risks. They reproduce quickly and evolve resistance to antibiotics we hurl at them far faster than we can come up with new, effective replacements. Medical researcher Elena Ivanova and her team have decided to take a different approach to combatting bacteria without relying on special chemicals, which the bacteria will invariably overcome. Wings of the clanger cicada (*Psaltoda claripennis*) are covered with a hexagonal array of tiny bumps, what Ivanova's team calls "nanopillars," and when a bacterium lands on these pillars, its cell stretches as it slides into a valley between the pillars, and ruptures if its membrane is soft enough. By virtue of its bumpy surface, the wing can destroy many incoming bacteria. The clanger cicada's wings also repel water, so even those bacteria that survive the hazardous bumps have a difficult time replicating in this dry environment. Medical implants, often colonized by bacteria, could stave off a dangerous film of bacteria if studded with clanger cicada–style wing surfaces.

Simply by mimicking the surface of an insect wing, we could create medical tools that self-sterilize.

Wings also pose relevant design solutions for bracing our injured bodies. Ali Khaheshi and engineering colleagues looked to an insect wing when faced with a seemingly intractable problem associated with splints and braces: How can you give strong support to an injured joint, but without entirely restricting the joint's motion? Dragonfly's wings have veins, and these veins have spiky joints that interlock so that movement of one part of the wing does not overextend its motion relative to other parts of the wing. A balance between strength and flexibility is essential during a dragonfly's aerobatic flights. Khaheshi's research team designed and 3D-printed a wrist splint that supports the joint while allowing a safe range of motion, using the same principle of interlocking components in the dragonfly wing to restrict hyperextension in a person's sprained or strained joints.

Insect wings can also inspire us to be on time. Before mobile phones gave us endless applications, including alarms, engineers were stumped when attempting to design a wristwatch with a mechanized chime. The challenge of creating an accurate timepiece so small, but with the power to generate a sufficiently loud alarm, seemed possible only after the physicist Paul Langevin pointed out that tiny noisemakers already exist in nature, in the form of singing crickets. Male crickets pull one wing, bearing a file, against the scraper on another wing, to produce a courtship call that can be heard by a female cricket from across a field. Engineers at Vulcain, in response to Langevin's comment, reproduced a cricket's chirping sound by fixing a peg to a membrane within the watch. Once a hammer strikes an anvil, the anvil resonates and is amplified through a chamber at a volume loud enough to alert the wearer. Thus was born a wristwatch alarm able to wake its wearer. To commemorate the natural source of inspiration, Vulcain released the Cricket Calibre in 1947.

Abdomen

Without our abdomen, we would have to relinquish our ability to digest food, excrete it, or reproduce. An insect's abdomen serves the same critical functions. The relatively soft posterior body region may seem less glamorous than an insect's head or thorax, and offer less potential to alter the course of human civilization, but to dismiss the abdomen would mean dismissing abdomen-generated products (honey, wax, venoms), the use of abdomens as human food, and abdominal features fueling biomimicry. We will end our biomimetic tour of the insect body with terminalia—those features at the posterior tip of the abdomen. Here, we explore the penis, egg-laying apparatus, and stinger, all in the name of medicine.

A common practice of getting something into or out of your body is to use a tube called a catheter. So versatile and effective, catheters date to ancient times across several civilizations, with the invention of the flexible catheter originating with statesman, scientist, and inventor Benjamin Franklin. Catheters need to be rigid enough to enter the body, but often flexible enough to snake around a convoluted path where it is needed. What in nature has evolved to accomplish both of these tasks? Why, the penis of the thistle tortoise beetle (*Cassida rubiginosa*)! (I should note that male arthropods' intromittent organs do not share a common ancestry with a mammal's penis, so an insect's penis is often graced with the obscure name "aedeagus.") Longer than the male beetle's entire body, his ten-millimeter-long penis winds through the coiled female reproductive tract before delivering his sperm to her eggs. To test the fortitude and flexibility of this natural wonder, a team of three scientists, including the very same Stanislav Gorb who brought us the submerged toy bulldozer to mimic the feet of a different tortoise beetle species, ran tests of penis stiffness and bending. Each test had to be conducted, I am sorry to report, on a severed member.

Only a few years after publishing their study, a team interested in preventing cerebral disorders invented a steerable microcatheter whose soft end was partly inspired by the tip of the thistle tortoise beetle's penis.

Transporting life-giving matter through a stiff but flexible tube is not the sole province of a single gender, however. The insect penis delivers seminal fluid with sperm, but the female's ovipositor transports and deposits eggs, sometimes in hard-to-reach, unusual places. Cicadas penetrate twigs, grasshoppers burrow into the soil, and parasitoid wasps pierce concealed animal

A pair of ichneumonid wasps seek hosts on which to lay their eggs. The top wasp is drilling into a tree with her ovipositor to lay an egg on a beetle grub feeding deep within as the second wasp feels for any vibrations caused by the chewing actions of other grubs.

hosts, all with ovipositors adapted to lay eggs where they stand the best chance of safely developing. Ovipositors have the ability to smell, which is important when you need to lay your eggs in a specific plant, or inside the right animal. It can be a drilling, steering tool, composed of multiple integrated parts that wind around obstacles to reach a specific target. From an engineering perspective, this is the perfect machinery to perform internal medical procedures. Marta Scali and a team at Delft University of Technology recognized this, and modeled a steerable needle after ovipositors of parasitoid wasps. Being just the kind of tool needed to navigate the delicate landscape of brain tissue, a variant on the ovipositor model allowed mechanical engineer Riccardo Secoli and his colleagues to conduct neurosurgery in 2022. They injected chemotherapy drugs directly into the brains of sheep, steering a robotic needle along a desired path.

Insects sting, not out of malice, but because they, too, want to live:
likewise our critics; they want, not to hurt us, but to take our blood.
—FRIEDRICH NIETZSCHE,
HUMAN, ALL TOO HUMAN (1879)

Our ultimate terminalia-inspired device will be familiar to anyone who has attempted to gain immunity to chicken pox, shingles, measles, mumps, rabies, a seasonal flu, yellow fever, hepatitis A or B, COVID-19, or other malady for which a vaccine exists. Can you guess which device has come to the rescue for so many, and which insect inspired its design? Consider the typical mode of delivery. You or someone you love has almost certainly been pricked by a hypodermic needle, and injected with a vaccine through a syringe connected to that needle. The history of injecting medicines is ancient, but the use of a true hypodermic syringe starts in 1853 with a Scottish physician named Alexander Wood. Wood coupled a needle with a glass syringe, which made it possible to view and deliver known quantities of liquid through a hollow needle with a sharp point into a patient. Wood's brother-in-law documented the physician's greatest medical contribution in a biography, making it clear what insect inspired the revolutionary device. When faced with a patient suffering from neuralgia (shooting pain caused by irritation or damage to a nerve), Wood knew that the patient's survival depended on sleep, and that administering a sleeping agent was essential.

A certain line of reasoning had led Dr. Wood to the belief that benefit was to be expected from the injection of morphia under the skin. Taking as his model the sting of the bee, he had constructed a small (glass) syringe, to which was attached a fine perforated needle point. This needle he passed under the skin, and through it he injected a small dose of morphia, which he could not give by the mouth. In this manner all derangement of stomach and liver was avoided, and immediate absorption of the morphia into the bloodstream took place. The strikingly beneficial result which followed this bold experiment made Dr. Wood aware that he now held in his hand a new method of treatment, which promised far-reaching results. Certainly in his most sanguine thoughts he could little have imagined, as he stood at that bedside, how in a few years every physician would be armed with that syringe, and countless patients would have seen cause to bless his skill.

Bless his skill, and bless the bee's stinger! Bees, wasps, and ants (the order Hymenoptera) often bear sharp stingers, typically coupled with glands that produce venom. For these insects, the stinger is a specially modified ovipositor. If a mother's egg needs to be inserted inside the body of a caterpillar or plant, having an organ that pierces and injects the egg in an optimal location is key. A little evolutionary tinkering, and an organ used to pierce and inject an egg can now pierce and inject venom, if you happen to be a stinger-bearing insect. Because the origin of their stinger is to lay eggs, don't expect male insects to sting, though males from several families of wasps use their spiny genitalia as "pseudo-stings" when threatened.

Evolutionary tinkering has only produced two other true stingers, and these two stingers look astonishingly similar, despite their belonging to very distantly related animals—scorpions . . . and a beetle. Insects use their antennae for sensing the world in different ways, but primarily to smell. Yet one beetle, a longhorn beetle from the Amazon (*Onychocerus albitarsis*), has a venom-injecting stinger at the tip of each antenna. The tip appears to sport a reservoir for venom with two pores emerging from channels through which venom flows. The beetle, if disturbed, jabs with antennae, leaving one of the scientists who described this evolutionary oddity in 2008 with a temporarily swollen finger.

If thoughts of hypodermic needles and stingers conjure only feelings of pain and fear, Seiji Aoyagi and mechanical engineering colleagues at Osaka's Kansai University may give you reason to embrace a family of insects that rarely receives positive attention. Mosquitoes are the infamous vectors of malaria, yellow fever, dengue, Zika, West Nile virus, Chikungunya, and more than their share of other diseases, and when not spreading historical scourges, can still annoy even the most patient and loving of people. But of the 3,802 described species, not all mosquitoes bite, and of those species that do, it is only the females, attempting to collect blood to nourish their unborn offspring. (Note: For this final stop on our tour of the insect body, we jump all the way back to where we began—the head.)

Whether we can admit it or not, we benefit from the existence of mosquitoes. They feed the masses—of fish, amphibians, birds, bats, and so many more. Ecological communities would crack and wither without mosquitoes. Even the despised bite of the mosquito carries surprising benefits for us.

Before we leap to her bite and blood meal, let's first imagine the gauntlet of dangers an expectant mother mosquito must face. There are predators to avoid, and swatting limbs to evade as she finds a suitable feeding site on which to land. Her two beating wings produce an audible whine. In his 1960 poem "Mosquito," wherein a human wishing to sleep waits for an opportunity to swat a noisy intruder, John Updike called this sound "a traitor to her camouflage." Updike's mosquito, like so many real mosquitoes, ends up a victim of murder.

For the lucky mosquito who lands without catastrophe, her success depends on sawing through tissue and drawing blood without rousing her host. A marvel of efficiency, a mosquito's proboscis delivers a near pain-free experience because she starts by administering an anesthetic, and the blood flows freely without clotting because she follows this numbing agent with an anticoagulant. Her stealth is exquisitely augmented by the mechanics of her mouthparts. By anchoring one pair of jagged maxillae and jiggling a tube (labrum) between this pair, she can probe around for a blood vessel without detection. The maxillae are shaped like notched harpoons, with the notches reducing the surface area in contact with our skin's nerve cells. We remain oblivious blood donors if our neurons aren't firing in response to a mosquito's touch.

She engages in another trick to minimize our detection of her blood-feeding. Her maxillae burrow through the skin in alternating motions to tap a blood vessel and her entire proboscis vibrates at thirty times per second while doing so. This action reduces the amount of force needed to penetrate, and the degree to which the area surrounding the bite is deformed. Less force and less distortion of surrounding tissue make a bite nearly imperceptible.

Aoyagi and other researchers have created three-part hypodermic needles that mimic the shape, the mechanical action, and the vibrating frequency of the mosquito's proboscis. Back in 2003, Aoyagi's colleagues tested themselves with their motorized, vibrating mosquito needles, and reported experiencing significantly less pain than that caused by a conventional hypodermic needle, though a dull pain lingered. The objective since 2003 has been obvious. A mosquito's proboscis is far more elaborate than a three-piece machine, involving additional mouthparts that help to steady the piercing, sawing, and sucking components. To achieve a pain-free injection means further perfecting a mosquito-inspired puncture.

Mimicking insect attributes can contribute to engineering triumphs. We began our foray into biomimicry with an exposé of our exoskeleton envy, but every facet of every insect has a story to tell or an engineering solution to admire. In the book *Biomimétisme: Quand la nature inspire la science* (2011), Mat Fournier collects examples of technologies inspired by nature, and of the sixty-seven animals and plants (and diatoms and kelp) featured, over one-quarter are arthropods, including fourteen insects (belonging to seven different orders), two spiders, and a lobster. This tiny sampling of arthropods inspiring inventions hints at the exponentially growing examples of insect biomimetics populating the relentless frontiers of technology. By the time I finish writing this, new technologies will have supplanted the old, and engineers will discover all-new answers from the behavior, anatomy, physiology, and genetics of previously ignored or maligned insects.

An eye here, a wing there, but what about building an entire mechanical insect? Biomimetics extends well beyond the piecemeal, and roboticists, engineers, and biologists have been making (crude) insect robots a reality for decades. Why insects and not robotic orangutans? Maybe the leap from observing insects to creating artificial versions of them isn't so daunting. It is tempting to watch a beetle ambling about and to see it as a wind-up toy or programmable machine, instead of as the behaviorally and physiologically complex animal it is. Allure of the insect robot manifests in robot toys and comic books featuring insectoid-humans or humanoid-insects, plated in armor and prepared for battle. Movie and television tropes of insect robots performing traumatic acts consistently paint a picture of swarms threatening human security, privacy, or our very existence. The foundations for such technologies have already been laid. Drones are named after flies, fireflies, dragonflies, and hornets. Cybernetic insect kits are commercially available. The age of the mechanized insect is upon us.

RISE OF THE
Insect Robots

A kitchen door opens, sending a gust of air toward a cockroach inspecting a morsel of food discarded on the floor. The movement of air particles deflects hair-like setae on a sensory appendage jutting from the cockroach's back end, and this stimulation sends a chemically driven message through the cockroach's nervous system. If the message were sent all the way to the brain and back to the legs, that could take too long for the rapid response required to make a life-saving dash below a cupboard. The message, instead, travels directly to the cockroach's leg closest to the stimulated appendage. Muscles contract in this leg, causing the leg to extend. Other legs follow in a standard set of alternating extensions and flexions, sending the cockroach scurrying in the direction opposite the source of the threat. Time after time, this sequence of actions spells safety and survival for the seemingly indomitable cockroach. Indomitable—until faced with an emerald jewel wasp (*Ampulex compressa*), a parasitoid with a gruesome, methodical means of transforming an agile escape artist into a compliant feast for her young.

First, the wasp delivers a venom to the base of the cockroach's front leg. This distraction temporarily paralyzes the cockroach's front legs and allows the wasp to more precisely insert her stinger to deliver a second venom. Using mechanical sensors on the stinger itself, she inserts the shaft into a special region of the cockroach's brain responsible for controlling the cockroach's escape response. The cockroach is not paralyzed. Indeed, the cockroach is perfectly capable of running as quickly as before the venom-laced violation, but does not. The wasp clips the cockroach's antennae, slurps up

208

hemolymph (an insect's version of blood) from the antennal stumps, then grabs the hapless zombie by what's left of one antenna and leads the cockroach into a dark burrow the wasp had dug in preparation for the acts to follow. Facing no backlash or hint of resistance, the wasp lays a single egg on the cockroach's middle leg. The tender maternal preparations end with a live burial, protecting her offspring and extending the shelf life of the host. A larva hatches from the egg, cuts open the leg on which she was born, laps up hemolymph for two of her larval stages, and on day six, as a third-stage larva, chews her way inside. The larva sanitizes her host with a bath of antimicrobial agents, and develops for weeks inside the still-living host, eating muscle and fat, followed by organs of choice, and ending with the central nervous system, all the time avoiding consuming the gut. The cockroach eventually goes "gentle into that good night," desiccated but never rotten, as the wasp finishes eating, spins a cocoon, and pupates within the husk of her host.

The cockroach and the wasp each exhibit a deliberate, predictable, and largely innate sequence of actions that can make them look a lot like miniature robots following simple sets of rules. Like a robot, an insect receives inputs (senses the world), processes the inputs, and produces outputs (responds). Even the most complicated animal behaviors can be broken down into simpler steps and examined mechanistically, but insects, being more alien to us than our backbone-bearing relatives, are easier to imagine as programmed or remote-controlled machines. That perception, plus the fact that insects can move efficiently into less accessible places, has made insects the go-to model for fashioning robots.

Engineers design robots that mimic cockroaches and other terrestrial insects because they so easily scamper over uneven surfaces and through cluttered terrain. Insects are also masters of scooting across water and, as early as 2003, artificial water striders have been imitating the aquatic acrobatic feats of their natural counterparts. Robostrider, built at MIT, was the first, and though it skates along with less elegance than its living model, *Gerris remigis*, it remains afloat without breaking the water's surface tension. Water striders are also able to leap from the water's surface, and a team at the

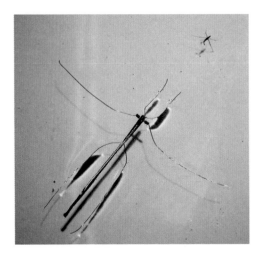

Harbin Institute of Technology in China, and another team at Seoul National University and the Wyss Institute at Harvard University, independently created robotic water striders that can do the same.

When mimicking nature, an engineer tries to find a simple solution to a difficult problem, and progress is often made inch by inch. I mean this literally, when that engineer attempts to simulate the dynamics of caterpillars. An inchworm stretches forward, grabs hold of a twig with front legs, and follows by drawing up their hind end, looping forward sometimes one inch at a time. This simple repeated motion is the hallmark of an entire family of moth larvae (Geometridae, appropriately derived from words meaning "Earth-measurer"), and is successful enough that this inching motion is shared by more than 23,500 species, and a bevy of mechanical doppelgängers. One inchworm robot, embedded with fiber-optic sensors, responds to real-time feedback as it scales walls. Air pumps through its silicone body as suction cups latch onto vertical surfaces. Another inchworm robot is propelled forward when electricity causes piezoelectric ceramic to bend, powering the body's motion while its asymmetric feet use friction to hold on to vertical surfaces. iCrawl is a third robot that uses electromagnetic feet specialized to scuttle along metal pipes.

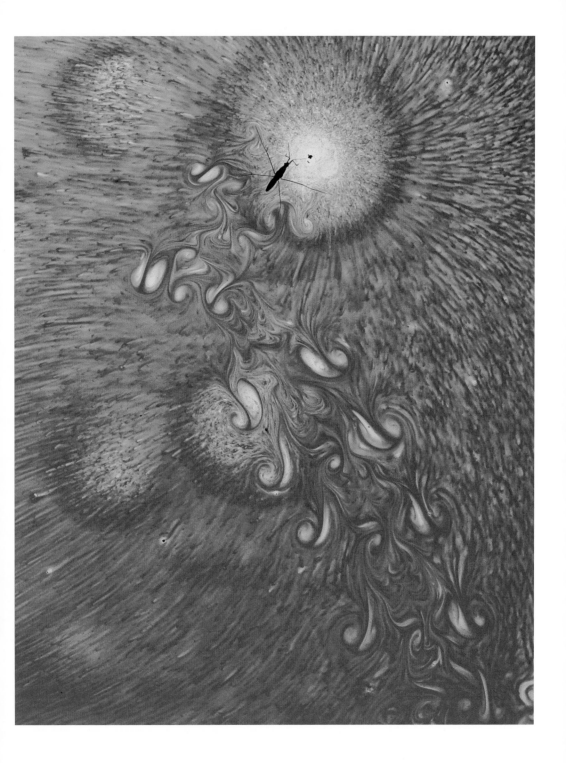

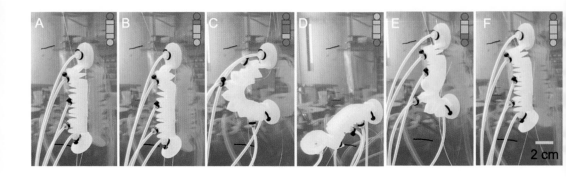

Inchworm robot, its silicone body pumped with air, latches to surfaces with suction cups.

If a single insect-inspired robot excites or alarms you, picture thousands operating without a leader, yet working triumphantly in unison. This type of decentralized control is largely how social insect colonies operate, successfully coordinating their swarming to new homes, foraging from profitable sources, cultivating disease-free gardens, or defending nests from massive invaders. To mimic what social insects do so effectively, engineers are creating leaderless, but collective, problem-solving robots. An ability to collectively solve problems is often called "swarm intelligence," and decisions made by robots given a few simple inputs can perform surprisingly complicated tasks.

Robots programmed to behave more like ant colonies—cooperatively and with a division of labor—brought more "food" back to their designated homesite. Other robots, when given options of paths to travel, made efficient choices much like those observed in trail-following ants. Robots programmed to interpret odors as either attractive or repellent organized themselves as social insects might. Robotics entrepreneur Rodney Brooks, who helped bring us the automated Roomba to vacuum floors while frightening pets, envisioned early on that legions of insect robots would become *Fast, Cheap & Out of Control*, as presented in Errol Morris's documentary of the same title. Brooks argues that inexpensive, autonomous robots unleashed en masse for planetary exploration would be more successful than relying on fewer, more expensive, ground-based robotic alternatives.

Like Brooks, many foresee insect robots solving problems with minimal computational effort, similar to that exerted by an insect. When Frances Chance—a computational neuroscientist at Sandia National Laboratories in New Mexico—sees dragonflies, she sees lessons in speed and efficiency. Chance believes that by better understanding how a dragonfly's brain works, we could produce models that work faster and with far less energy to solve problems we face that aren't entirely dissimilar to the daily tasks performed by dragonflies. For instance, instead of using brute force to fly after an insect along the insect's flight path and reacting to the insect's every move, a dragonfly predicts where the insect is headed. The dragonfly anticipates a prey item's flight trajectory and intercepts, and does so with astounding success (with as high as a 97 percent capture rate). This involves responding to flying prey within thousandths of a second by coordinating body segments and individually flapping wings at precise frequencies, all while tracking movement using visual resolution that is magnitudes lower than our own. As Chance quantifies the elegant efficiency of this task, she speculates about applications to improve artificial intelligence systems to intercept missiles or control self-driving cars. Chance would like to build a computer chip that not only accomplishes what a brain does, but in the same way as the brain does it.

Francis Crick, co-discoverer of the structure of DNA, would be pleased with this approach. In *What Mad Pursuit*, Crick made it clear that too many models of the brain are, what he called, "don't worry" theories, "demonstrations" that do not pretend to approximate reality, but merely produce "pretty" results. Efforts are growing to assign insect robots controls modeled after the activities of real insect nervous systems. For example, Case Western Reserve University's MantisBot, designed to use legs in multiple ways, is endowed with a visual tracking system that simulates neural circuits within a praying mantis. Don't mind the fact that the robot looks like a menacing green killing machine.

Ultimately, imitating the biology of the brain means mapping the brain, with all of its cells and their connections. For many brain scientists, building this roadmap—a "connectome"—would be a major step to understanding how a brain works. As I write this, only four species' brains have been fully mapped, and the most advanced brain map now belongs to the larva of a fruit fly (*Drosophila melanogaster*; the others: a roundworm, a larval sea squirt, and a larval segmented worm). This fly's connectome, with its 3,016 neurons and 548,000 connections, will someday help scientists uncover the neural basis of behaviors, including our own, and may lead to new brain-like machine-learning approaches.

Each insect-inspired technology comes with the promise of improving, or at least changing, our lives. Sometimes the potential applications are not immediately clear, but sometimes they couldn't be more obvious. Hollywood has taught us that robotic insects would, first and foremost, make great spies. This makes sense. Think about what a human spy would most like to become—a fly on the wall. To transform into a fly would mean slipping undetected into private quarters, and eavesdropping on secret conversations. Such a human-fly transmutation has not always worked out so well, however. Typically, the transformation involves teleportation technologies that result in terrifying human-fly hybrids (scientist Andre Delambre in George Langelaan's short story "The Fly," adapted for *The Fly* movie franchises). Alternatively, it can rely on postmortem reincarnation (Jacob Cerf in Rebecca Miller's novel *Jacob's Folly*, and Nani from Telugu and Tamil films *Eega* and *Naan Ee*, respectively. A safer bet is, instead, to control an insect to do one's bidding, or to create one from scratch.

Veiled in secrecy, a watchmaker and an amateur entomologist responded to the threat of the Cold War by helping devise what would become the smallest airborne robot in history. It was the 1970s and the obsession to collect intelligence drove unprecedented innovations in espionage technology. The US Central Intelligence Agency (CIA) had developed a tiny listening device but no clever way of covertly transporting it. The CIA has ferried or housed spy machinery via relatively gargantuan animals—pigeons, dolphins, an eagle drone, robotic catfish, and a distractible cat. Even their fake tiger excrement was excessively large, and would have looked suspicious if designed to move. What the CIA needed was a flying insect to infiltrate without rousing suspicion. Charles Adkins led the "Insectothopter" project,

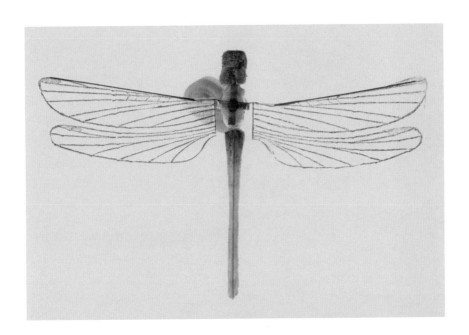

CIA's Insectothopter (prototype and operational model), built to fly and to spy

which at first was to feature a bumblebee, but the aerial dynamics of bumblebees were deemed too complicated, so a dragonfly enthusiast on the team suggested a more feasible candidate. The resulting hand-painted dragonfly robot has eyes made of glass beads. If sound vibrations (from a top-secret conversation, say) struck the glass, distorting it, a laser beam directed at the beads would travel a different distance back to a receiver. By analyzing the reflected laser beams, the sounds that caused the distortions in the beads could be recreated. This solution to eavesdropping via an optical microphone had some precedent, but the primary problem remained—how to fly the dragonfly spy.

Ultimately, the Insectothopter was propelled by a tiny gasoline engine as fake wings flapped up and down and excess gas was vented backward, providing additional thrust as the faux insect flew. Shining a laser beam on a bimetallic strip dangling from the abdomen would open the vent and effectively serve as a rudder. Though capable of flying for sixty seconds and a total of just over 219 yards (200 m), the harsh reality of field conditions doomed the project. The one-gram robot could be blown off course with a slight crosswind, and manually training a laser on such a moving target to redirect it was too difficult. The Insectothopter was shelved and remained unknown to the rest of the world until 2003 (except the former KGB, who, according to the International Spy Museum in Washington, DC, fashioned their own dragonfly Insectothopter in the 1990s), and more fully when the original files were released in 2019. You can visit this early insect robot relic in the CIA Museum, should you have clearance to CIA Headquarters. For those of us who do not, still images and videos immortalizing the ephemeral flight of this insect robot pioneer are available on the CIA website and all over the internet.

This was far from the first time insects had been used to inspire acts of espionage—or torture or war for that matter, and it won't be the last. The race is on to produce the first of a fleet of robots that can perform with the power and ability to smoothly traverse treacherous territory with infrared imaging and global positioning capabilities, sensors for detecting chemicals and radiation, and an ability to communicate. Research labs at Georgia Tech,

MIT, Harvard, City University of Hong Kong, Delft University of Technology, South Korea's Konkuk University, University of Tokyo, the company AeroVironment, and many others are tackling a range of challenges in a quickly evolving, competitive arena of robotic insects.

The most ambitious challenge is to devise smaller and smaller flying robots, many of which are, not coincidentally, clear mimics of a diversity of insects. The reason for mimicking a variety of insects is that there are different ways to fly and each serves its own function. Some insects flit erratically, while others hover, weave, or dart. Flying insects also vary in size, from parasitoid wasps smaller than an amoeba that effectively swim through the air, to a Hercules moth (*Coscinocera hercules*) that has more wing surface area than a blue jay. The diversity of insects selected as models is often apparent in the robots' names: RoboBee, RoboFly, DelFly, Wasp AE, Artificial Butterfly, and Dragonfly, to name a few. Since the Insectothopter, insect robots have grown (as they've shrunk) more sophisticated. They can survive collisions, are soft and crushable, carry minute solar cells, are autonomous, have wings that flap nearly five hundred times per second, can hover, or can somersault in the air. With every design come trade-offs, so no single robot presently exhibits all of these features.

As promising as insect robots are, no flying machine has the benefit of hundreds of millions of years of evolution to fine tune its mechanics. Natural selection hasn't operated on robot sensors, processing machinery, and mechanical outputs, so robots lack the elegance, finesse, and efficiency displayed by their living counterparts. Power sources for robots are heavy and onboard controllers are simple, whereas insects generate their own power and their brains are remarkably sophisticated. Since nothing can presently match the flight capabilities of a winged insect, some engineers have accommodated their designs to hijack actual insects. Their solution: Use insects, but attempt to guide their flight. The following interventions were made with living insects, and come with troubling ethical implications that will become clearer and more obvious as we learn more about insect biology.

In 2009, Siva Pulla and Amit Lal used dental wax to glue backpacks containing motors, propellers, and a helium balloon to a pair of tobacco

hornworm moths, and guided the moths' flight for 65.6 yards (60 m). To more effectively (and invasively) steer insects, others have inserted electrodes through an adult's exoskeleton into nerve cells, chosen after careful observation and understanding of the insect's behavior. If a beetle's response to flight depends on light level, you can target neurons where light is first processed by the insect's brain. Send electrical pulses to this region of the brain and the beetle will take off or cease flying, as desired. If a moth rotates her head in the intended direction of flight in anticipation of flying in that direction, you can stimulate the appropriate neck muscle and elicit a left or right turn. The extent to which beetles, moths, grasshoppers, cockroaches, and dragonflies have been modified and mechanized has been constrained only by the size of the insect and the insect's ability to lug the technological burden, and is no longer exclusively the domain of fancy research labs harboring prohibitively expensive equipment. You can purchase or design your very own RoboRoach Bundle, complete with a printed circuit board you glue to a cockroach's back, electrodes to connect to the antennae, a lithium battery, and LEDs to show components are working. Presto! You can then control left and right turns wirelessly using a phone app. Stripping a cockroach of this bit of free will is ephemeral, however, as the cockroach's brain wrenches back control within several minutes.

There are ways to control movement as well as the *mind* that do not require electrodes, wires, or any cybernetic components. The key to controlling an animal's motion is to more fundamentally circumvent their nervous system. Luigi Galvani demonstrated that an animal's movement is governed by electricity generated by the animal itself. In the late 1700s, Galvani caused a frog's leg to twitch, first by applying electricity to a frog leg in his laboratory, then by dramatically hooking a frog leg up to a lightning rod during a storm. The shock of reanimating a corpse rippled well beyond science circles. Mary Wollstonecraft Shelley, on the night following discussions of "galvanism" and other science topics between her husband and Lord Byron, conjured up the beginnings of *Frankenstein*, in which Victor Frankenstein ventured to "bestow animation upon lifeless matter," only to reject his creation as an abomination—a "vile insect."

Knowing that nerve cells communicate electrically, and that the passage of chemicals from one nerve cell to the next through pores is necessary to transmit the electricity, Gero Miesenböck and colleagues genetically engineer fruit flies so that the nerve cells' pores are coupled with light-sensitive pigment molecules. Should they wish for a fly to take off or land, they simply flash a light, the pores open and the nerve cells fire. The light that flashes could shine from a portable unit affixed to the insect's back, if the insect is large enough.

The beauty and terror that could arise from tinkering with flying insects is only partly foreseeable. That is the way with basic research. Techniques developed to stimulate flight in insects could culminate in designing better robots, orchestrating nefarious acts of cybernetic warfare, or transforming our approach to confronting the threat of neural diseases through deep-brain stimulation.

Equally lofty is the aim of understanding how Earth's oldest and most diverse flying organisms maneuver so gracefully on the wing. Dragonflies, for example, can move each wing independently, hover, fly backward, and capture prey like no other predator. Jesse Wheeler, biomedical engineer, and Anthony Leonardo, team leader at Janelia Research Campus, led such a project in 2017, creating cyborg dragonflies. DragonflEye could record and transmit data from the inner workings of free-flying dragonflies. Unlike the Insectothopter, DragonflEye was a living dragonfly transmitting clues that could help us understand the mechanics, evolution, and behavior of one of evolution's greatest aerial success stories.

Art,
AND FABRICATING THE
PERFECT INSECT

Years ago, I worked (and secretly lived) at the American Museum of
Natural History, making exhibits by day and roaming its half-lit halls by
night. Before this, I had spent time making models for museums at a natural
history display–making studio in the Ozarks. I joined a crew of top-notch
artists, hidden from the rest of the world in the Missouri woods. A welder,
carpenter, graphic designers, mold-makers, painters, and sculptors spent
their days together, creating spaces that would transport future museumgoers
to forests populated by extinct giants, among fish in deep-sea Devonian
waters, or confronted with magnificently magnified minutiae within a drop
of pond water. I arrived at Chase Studio with little sculpting experience, and
left with the peculiar aptitude for fabricating realistic insects—sufficiently
realistic that museum visitors would pass them by with the same attention
they might pay to an actual insect, which was often none. The Exhibition
Department at the American Museum of Natural History was hiring, and
I packed my things and moved from a condemned trailer in Mark Twain
National Forest to an illegal sublet in the Bronx, and found myself creating
African arthropods and giant viruses in New York City. I helped craft models
for the Hall of Biodiversity, and was, once again, immersed in the obscure art
of making insects. One evening, as I was preparing for another night alone in
the museum and my last remaining colleague was running out the door, she
urged me to attend an art exhibit and dropped a folder full of photographs
in my hands. I flipped through black-and-white images of immature aquatic

insects. All seemed oddly normal until I found a photograph with a shocking reveal: a fish hook protruded from one of the insect's undersides. These were fakes, and I—an entomologist and a fabricator of insects—had been hoodwinked.

Bill Logan, as artist and angler, invents ways of creating insects. His primary objective is to fool fish into preying on his hooked counterfeits. This recreational practice may date back two thousand years, and involves a few basic steps. First, he ties a lightweight "fly" with a hook firmly in place, attaches the artificial insect to a line, then flings the lure into waters likely to house finned predators. As a fly-fishing fanatic (he once divulged to me a top-secret handshake restricted to the hardcore among his brethren), Logan creates his insects using a mix of traditional and modern methods and materials.

Beginning in the 1990s, Logan's desire to tie the perfect fly escalated far beyond the conventional and the utilitarian, resulting in flies hailed for their hyperrealism and exhibited as pieces of art. No bait in history can match Logan's flies. He has invested more than one hundred hours when tying a single fly that could easily blend into a natural body of water, or pass as a specimen in an insect collection. When aiming for perfection, Logan would intentionally include imperfection. All too often in nature, an insect suffers a broken antenna or cercus (posterior feeler)—a sign of an individual's experience, escape, or developmental deformity; or the mark of a Logan fly.

Logan's materials include bleached-browned peacock feathers, ostrich feathers, porcupine quills, insect pins, thread, monofilament, polymer clay, epoxy (never used as an adhesive), brass shim, hand-sanded copper wire, paint, varnish, the mysterious and alluring-sounding Magic Shining Foil, and a modified fish hook. Though fashioning human flesh out of stone is marked as the pinnacle of art (think Michelangelo's *David*), technically audacious, beautiful recreations of nature come on much smaller scales, and Logan's fly ties join an eclectic

Hyperrealistic tied fly of a western golden stonefly nymph (*Hesperoperla pacifica*) and pages from Bill Logan's notebooks: western golden stonefly and eastern green drake mayfly nymph (*Ephemera guttulata*). Every anatomical detail, every observation is captured in Logan's richly annotated, personal books.

Maxilla and **Labia structure**

paraglossa made the same way as in eastern. palps (labial) also the same made and mounted.

Maxilla assembly

① stainless steel brush wire bent like SD, width of head at widest point

final jaws a bit longer than mandibles

② 3 copper wires laid in position and prepared (cut on mandibles) easiest way to mount is to put each on w/2 lashes and rough position then back on lash off + mount next w/2 and same procedure. Also mounted palp. a couple of more wraps to back then position, clip etc using car. pers for sizing.

③ palp slightly larger — each towards see it a bit shorter than last.

WELL cemented

④ X wrapped + twisted + into position. filed flat each side

Also wrapped around once or twice

Once maxilla is mounted re- tips and then epoxy them

Epoxy

① coated attachment and base of maxilla, applying epoxy. mention paraglossa before applying epoxy.

also gap on underside at head base, tried to get some attachment to underside of mandibles

cross-section

looped around ponch then under between hook and shim. Loope in front of palps.

② underside of labrum and coat on 1st maxilla. thin coat, flattened underside w/wedge stick 2nd round for second maxilla and touch up

ventral view wrapped blind

underside of head (submentum mentum)

• underside of head see pattern in file
→ See pattern in file → straight epoxy
ⓐ cutout pattern in straight epoxy (w/ #6) sheet — no coroid strip
ⓑ laid in and tied down at

Ventral side, meta, mesolegs —

④ dirty, slightly b. sienna reddish cream legs much lighter cream on femur and tibia. Fairly heavy but smooth as possible drybrushing w/ straight paint. * if possible — legs of this on meso legs so they are a touch darker.

OF SPECIAL NOTE — LEG INJURY on MESO leg — felbrew painted w/rust or rust. Then: Same procedure as above

Then: Dark edge (like cox area above) Then small dabs of hey-cream color- heavier than dry-brushed areas and was lighter - also trimmed these dots back w/solvent brush to settle them in a bit.

④ → left ventral prolegs as they were.
⑤ → DORSAL STARTS HERE

Ab segs done as follows:
ⓐ: darkened A10 w/ a bit of b. sienna and some of same w/a bit of yellow like this: first lighter color in a line along tbutg edge darker color in a line dry brushed on then (behind thread wraps, visible as a slightly lighter line under epoxy at front of segms)

---

cross of gill tufts, which are done like this:

brown black moderately thin applied and edges softened w/solvent brush - ckn lot → it go slightly up into ostrich stalks

Then, thicker darker spot in middles and tap- softened so they don't look like paint dabs.
* that's it for Ab
⑥ → yikes Thorax time!

③ - dots + dashes of light and dark side NEXT PAGE

keep pattern and low

① brown black- drybrushed along edge- Don't go around under wingcase contour
② Edging + suggested patterning each photo + spec. I have lighter so I just took the average - black-brown heavier - worked back into a pattern w/ both in + sharpened toothpick + p
③ b. sienna + yellow

assortment of grand attempts at creating miniature masterpieces. People have built insects out of paper, bamboo, leather, flowers, feathers, fabric, felt, glass, epoxies, metal, and, like Logan's fly ties, an array of mixed media.

Why create a hyperrealistic, artificial insect, especially if it will never be used to catch fish? Some are made for displaying, some as toys for playing, and others for playing pranks. All you have to do is read reviews of products online to know that plastic cockroaches make for wholesome, heart-stopping fun. Fabricating insects can be a traditional practice, a technical challenge, a way to creatively evoke emotions, or an engineering feat with applications. We have thought about how real insects are used by artists, scientists, and others, but real insects come with limitations. Living insects are difficult to control in complex ways, and they die. Dead insects dry up and their colors usually fade. They are brittle, and attract beetles, which quickly turn the corpses into piles of dust. An artificial insect often not only lasts longer, but can be created with exaggerated sizes, colors, or structures, as befits an artist's vision, an educator's teaching lesson, or a scientist's experiment.

Art and the Artificial Insect

Pharaohs of ancient Egypt were deemed all powerful, and with power, came insects. Starting in the First Dynasty (circa 3100–2900 BCE), rulers of Egypt adopted a throne name "of the Sedge and Bee." The hieroglyph of a sedge (symbol of Upper Egypt) was coupled with the hieroglyph of a bee (symbol of Lower Egypt), the bee being associated with honey, wax, and wealth. Scarab beetles, rolling their balls of dung, also demanded the respect of the Egyptian kings. Dung-rollers were linked with Khepri, god of the rising sun. Sacred scarabs were carved out of glass, bone, ivory, and precious metals, and became popular fashion accessories. The allure of wearing artificial insects has never died, but has changed alongside fashion trends, with glorious examples rising out of the Art Nouveau movement. René Lalique (1860–1945) created pieces for Maison de l'Art Nouveau, a shop in Paris from which the name of the movement originated. He showcased scarabs in a corsage of gold and enamel, wasps in a hatpin, and dragonflies in a pendant. Exhibiting power and beauty, Lalique's tour de force is the "Dragonfly-woman," with

Egyptian scarabs from the Middle Kingdom (circa 1850–1640 BCE) and the New Kingdom (circa 1492–1473 BCE). Hieroglyphs are of bee, symbol of Lower Egypt, and sedge plant, symbol of Upper Egypt.

BOTTOM: "Dragonfly-woman" corsage ornament by René Lalique (gold, enamel, chrysoprase, chalcedony, moonstones, and diamonds [1897–1898])

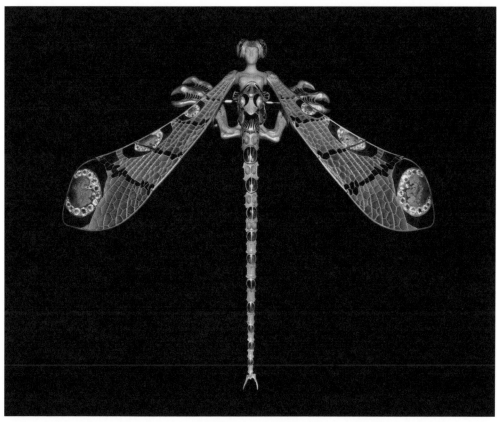

articulated, open wings, griffin claws, double-scarab beetle helmet, and bust of a human.

Metal is a popular choice among insect fabricators, probably because the insects themselves are a bit like knights in armor. A bronze mantid by University of Chicago biochemist, Allan Drummond, is held together with neodymium magnets. Open a violin beetle of silver and gold by Nova Scotia–based metalsmith Elizabeth Goluch to reveal a hidden metal bow. These, and any other lifelike works in metal, owe a debt to jizai okimono, the traditional Japanese practice of creating authentically articulated animals. The oldest known example is a dragon by Myochin Muneaki (1713), and the oldest insect is a butterfly by Myochin Muneyasu (1753). The Myochins built samurai armor until the need for war gear dried up during peacetime. Applying their talents to decorative metalwork instead, they created animals (real and mythical), and for a time, jizai okimono was profitable. Insects were a popular subject, and the tradition of fabricating them has continued quietly ever since. Only one such master remains in Japan. Haruo Mitsuta had thought the practice extinct for a century until he discovered a fifth-generation jizai craftsman, and convinced him to be his mentor. Mitsuta carries on the jizai legacy, crafting authentically articulated arthropods, first by carefully examining specimens, drawing blueprints of their bodies, dismantling them, and studying each and every anatomical feature as he recreates them in metal. Cutting, hammering, filing, grinding, polishing, adding springs so mandibles faithfully open and close, soaking in a darkening chemical bath, and blasting with a flame torch are all part of the process. Legs bend, mandibles pinch, and wings unfold—and each realistic component was created in metal by hand.

Move legs and body parts in Ricky Boscarino's articulated *Hieronymus Beetle* or open Elizabeth Goluch's *Violin Beetle* to find a tiny bow made of silver and gold, modeled after the violin beetle, *Mormolyce phyllodes*.

As materials differ, challenges of building anatomically accurate insects can also differ. Tools and techniques vary, but the need to deeply understand the
subjects is universal across artists daring to produce replicas of nature. Actual-size insects strike dramatic poses
in bamboo (Noriyuki Saitoh), or stand as giants in wood
(Patrick Bremer). Fabric moths are sewn (Annemieke
Mein) and artificial butterflies flap in jars in response
to sound (António Caramelo). Leather beetle pouches
and flea bags are made to carry belongings (Amanojaku
to Hesomagari), and mechanical insects can be hand-cranked to flap wings
(David Beck). Some artificial insects are also meant to be eaten. Copenha-
gen's Noma, frequently hailed as the world's best restaurant, is a place where
culinary innovation and esthetics are valued over portion size, and a recent
edible flourish featured intricately crafted fruit leather beetles.

Creating highly realistic insects—to be worn, to be eaten, or simply to be experienced—is in many ways like any other art. Insects evoke different emotions in different people, and artificial insects can inspire, amuse, terrify, or enlighten. My father, Arnold Klein, co-owned my parents' art gallery in Michigan for forty years, and knew more about art than anyone I've ever known. Many years ago, I pressed him to give me a definition of art, and without hesitation he replied, "It's art if it's not anything else . . . unless of course it's a bicycle wheel or a urinal." Art is difficult to define, but it can take all forms, and it can be motivated by its creator in as many ways as art can affect its audience.

Educating with Artifice

I can remember the fear and apprehension I felt when we opened the glass case. There he was, glorious and perfect, save for one glaring blemish. Millions had admired him, and millions more to come would gape at his beauty and gawk at his abnormal dimensions. Seventy-five times the diameter and four hundred thousand times the volume of the actual animal, the model of a male mosquito was made well over a century ago, by and under the direction of Bror Eric Dahlgren, curator of Arts and Preparation at the American Museum of Natural History. It was later featured as part of a larger exhibit about insects and the spread of disease, which feels odd to me given that male mosquitoes cannot suck blood and never vector disease. Only by association (his female counterpart can carry a single-celled *Plasmodium* that causes malaria) does this *Anopheles maculipennis* mosquito warrant any misgivings.

A piece had fallen from his head and the giant was broken, possibly the result of one too many visitors banging the case or tapping the glass. My job was to replace the detached bit. As a preparator at the museum, I had built piles of insect models, and repaired public displays in several of its halls. I once sculpted and painted a poison dart frog to replace one stolen from the Hall of Biodiversity's Spectrum of Life. I glued a new claw on the foot of the rotunda's *Allosaurus* so he could continue to dodge the fury of the *Barosaurus* skeleton rearing up to protect her offspring. But this was different. This was the most elaborate and realistic mosquito model in the

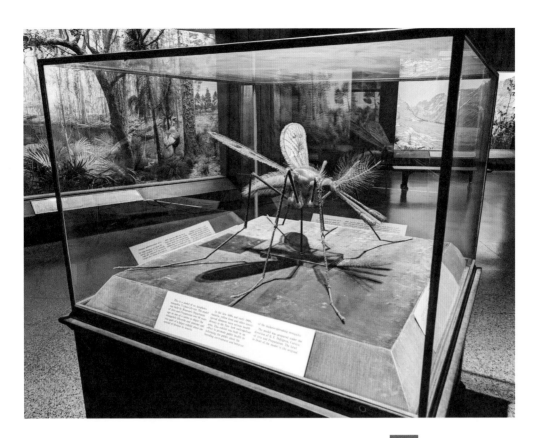

world, a unique piece of art meticulously prepared in the early days of the American Museum. It was highlighted as one of the museum's one hundred treasures. I exhaled. I took my time carefully reattaching the piece with wax and attempting to match the original grace of Dahlgren's model. When I finished, we enclosed the treasure for future visitors to admire. That was twenty-five years

A male *Anopheles* mosquito created 75x actual size by Bror Eric Dahlgren more than one hundred years ago to educate people about malaria

ago, and the model of the male mosquito now sits alone in the middle of an exhibit hall. Every time I return, I flinch when faced with the restoration. I hope I am the only one who notices. Better for me alone to be haunted by the imperfection than the piece be diminished in others' eyes.

Art can educate, and some of the most exquisitely realistic artificial insects are built to teach. Like the giant mosquito, sculptures sit behind glass or serve as demonstration models in natural history museums to offer

lessons about anatomy, physiology, behavior, ecology, or evolution. Model makers, like Lorenzo Possenti in Italy, specialize in fabricating giant insects for museum exhibits. As I write this, Chase Studio, nestled in the Missouri Ozarks (where I got my start creating insects for museums) is celebrating its fiftieth anniversary of creating ancient and contemporary scenes to teach museumgoers about the natural world. Chase Studio insects are made with urethanes, adhesives, metal wire, or other relatively sturdy materials, especially if the models are meant to be touched.

Rewind two centuries and instructional models of insects were made of papier-mâché. Unlatch a hook and open the body of a beetle to expose its labeled paper anatomy, compliments of Louis Thomas Jérôme Auzoux (1797–1880). Auzoux was inspired by a visit to a papier-mâché studio in 1820, and applied techniques of puppet and doll making to found a business creating models for instruction.

A century after Auzoux's industrial and industrious efforts to educate with models, a preparator working for the Museum für Naturkunde in Berlin

Instructional models made of papier-mâché: a honey bee drone and cockchafer beetle (Louis Thomas Jérôme Auzoux; bee circa 1875, beetle 1878)

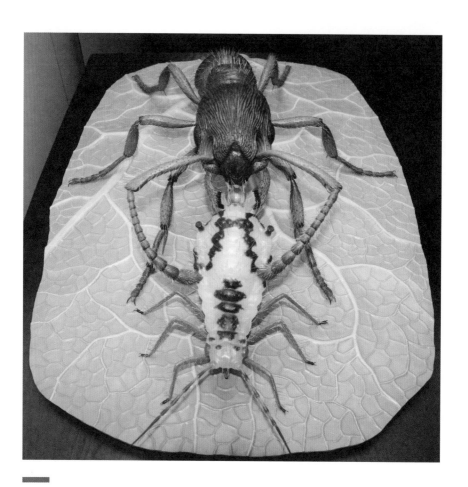

A maple aphid (*Drepanosiphum* sp.) appeases a common red ant (*Myrmica rubra*) by excreting a sugary drop of honeydew, as modeled in papier-mâché by Alfred Keller (1940s).

created a series of educational insect masterpieces, also with papier-mâché. Investing one year per model between 1930 and 1955, Alfred Keller constructed giant insects in plasticine, cast in plaster as a reference, and recast in papier-mâché, adding details with wax. A fly is covered in 2,653 bristles made of galalith (synthetic plastic combining casein with formaldehyde), and wings of celluloid rest naturally and transparently to either side. A mosquito is poised in flight, a wheat weevil nibbles on a kernel, and an aphid appeases an ant with a drop of honeydew. Bombs dropped during World War II obliterated some of Keller's models, but glorious examples remain.

Another unlikely material for educational models was made famous by a Czech father and son team based in Germany. For over fifty years (1886 to 1936), Leopold and Rudolf Blaschka produced thousands of organisms out of glass, 4,300 of which fill Harvard Museum of Natural History's "Glass Flowers" collection. In it, you can find glass insects pollinating glass flowers. These instructional models are more intricate and accurate than Auzoux's papier-mâché predecessors, but they are also infinitely more fragile.

Among thousands of glass models of flowers at Harvard, life-size glass bees pollinate *Rossioglossum grande*, and a single enlarged bee pollinates a magnified flower.

Models (other than by Keller or the Blaschkas) are passed around in classrooms, and can serve as hands-on tools for the visually impaired. Insect dummies also save insect lives as alternatives to classroom dissections. Though dissecting real insects can hone one's manual skills and reveal variation across individuals, a model often does a better job of conveying information than the real thing. This is because preserved specimens are often shriveled, rubbery, and contain monochromatically gray innards, whereas models can be enlarged, naturally textured, and realistically colored, highlighting features of interest. Though digital dissections and computer models are becoming more and more common, and dioramas are disappearing from museums, tangible models, including 3D-printed or hand-sculpted insects, will continue to have a place in education.

Science and the Simulated Insect

Fabricating insects is not a practice limited to art or education, as we learned with studies involving insect robots. Sometimes creating a convincing insect model provides a vehicle for discovering how nature operates. Physical models expand a scientist's toolbox, and make otherwise impossible questions testable. You can control the look, the placement, and the behavior of models in ways that might not work with the real animal. Try making frogs call without moving their vocal sacs, a fish lead a school in a desired direction, or an insect display colors they do not have in nature, all in carefully controlled, repeatable ways.

Biologists Dan Papaj and Ginny Newsom at the University of Arizona wanted to know if pipevine swallowtail butterfly caterpillars not only deter predators by advertising their toxicity using bright coloration, but also benefit from the colors by deterring other pipevine swallowtails from laying eggs in their vicinity—which would lead to competition. To test whether natural color patterns alone did the job of warding off the would-be egg-layers, Papaj and Newsom made silicone molds of real caterpillars and cast hot-melt glue replicas, then painted them. Sure enough, butterflies laid fewer eggs near the naturally painted models versus models painted green or when no models were present.

The same year, Papaj worked with his postdoctoral student, Brad Worden, to find out if bumblebees learn to forage on certain flowers by watching other bumblebees. They exposed their bumblebees to either living bumblebees from a different colony or artificial bees, both near flowers of different colors. If their test bees saw other bees (live or artificial) near green flowers, the observing bees then preferred green flowers themselves. Had fake caterpillars and bumblebees been excluded from the two studies, the researchers would have had no idea if other factors, like smell, had caused the butterflies' and bumblebees' behaviors.

Fake caterpillars and bumblebees aren't the only models that have duped real animals in the name of science. For additional examples, biologists relied on insect doppelgängers to address the following questions.

Q1: Can a bat find an insect sitting entirely motionless in a cluttered landscape?

Technique: Make paper or aluminum dragonflies with smooth wings, aluminum dragonflies with crumply wings, and present *Micronycteris microtis* bats with either these, or combinations of unaltered or altered dead dragonflies.

Result: The bats not only pick off motionless prey sitting on a leaf, but they can also distinguish shape, surface structure, and material of their prey using echolocation. (They preferred aluminum dragonflies with crumply wings over fake dragonflies with smooth wings.)

Q2: How do male butterflies distinguish females of their own species from females of a very closely related species?

Technique: Digitally modify pictures of *Lycaeides melissa* wings on paper dummies and present the dummies to real *Lycaeides idas* butterflies.

Result: *L. idas* males use subtle pattern differences on the wings when distinguishing among mates. (*L. idas* males chose the paper models with smaller spots, resembling their own females.)

Q3: Do jewel beetles attract mates using iridescence caused by the surface structure of their exoskeletons?

Technique: Create decoys with "bioreplicated" structural detailing, 3D-print a beetle without, and paint both green. Present these and other decoys to *Agrilus biguttatus* beetles.

Result: Males use the iridescence caused by fine surface structure details when selecting mates. (Beetles flew and landed on the bioreplicated decoys more than the 3D-printed alternative.) Here, the iridescent decoys can shine a light on mate selection in a species known to damage oak trees, and advance technologies related to material science. In all of these examples, artifice gives us a clearer window through which to test and understand what is real.

Architecture:
BUILDING WHAT THEY BUILD

Eugene Tssui, architect, fashion designer, author, and eight-time champion amateur boxer based in Emeryville, California, wants to change the world. His architecture is designed to shout about the climate crisis, and serve as a contrast to the wasteful, toxic, and outdated structures in which we live and work. Critiqued, dismissed, and ejected from architecture programs, Tssui has more recently been embraced as a pioneer and visionary of biologically inspired architecture. He is the subject of three documentaries, has been featured in a MoMA retrospective exhibit, and has been asked to speak at both the United Nations San Francisco branch and the United Nations General Council in New York City.

Intrigued by his vision and his dramatic persona, I wrote to Tssui. Not only did he respond immediately, but we talked into the night about insects, architecture, sustainability, martial arts, and how people need to follow their passions in life. Tssui reveres insects, and it clearly shows in his works. His parents live in an earthquake-proof house modeled after forms found in the most indestructible animal, a close relative of arthropods called a tardigrade (Tardigrade/Fish House, Berkeley, California, 1995). A roof opens with hand-cranked dragonfly wings (Reyes Residence, Oakland, California, 1993). The Watsu Health and Medical Center (Middleton, California) uses the spherical shapes of honeypot ants and oak galls. (One caste of honeypot workers accepts sweet food from nestmates until their bodies are bloated like grapes, hanging from the ceiling of underground chambers, ready to feed their sisters during times of need; oak galls take a similar shape

Walter Tschinkel's aluminum cast of a nest of fire ants (*Solenopsis invicta*)

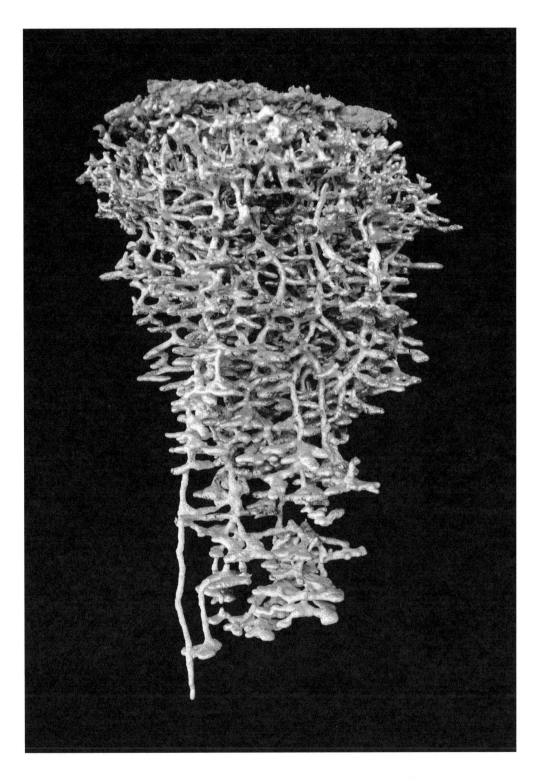

when oak trees are chemically manipulated by wasps developing within their tissues.)

Tssui's book *Evolutionary Architecture: Nature as a Basis for Design* draws from insects and other nonhuman animals to produce innovative and imaginative designs for human architecture (ride in the "Cicada" airlift photovoltaic mobile home), but it also explores the biology of organisms, allowing the reader to imagine possible lessons we could learn from untapped architectural success stories all around us. In a section on "natural forms," Tssui writes about the biology of one frog, two fish, two mammals (including the beaver, of course), ten birds, and a dozen categories of arthropods. Architecture could benefit from the triumphs of arthropod nests, or silky webs, cases, cocoons, and nets.

I asked Tssui what insects have to offer his architectural work, and he responded with the following:

> *Before I begin any architecture project, I find out what are the governmental limitations of the site and its unique characteristics and what are the requirements and aspirations of the owners. Then I ask, what organisms live on this site? This includes the flora and fauna, the soil, the insects, the birds and reptiles and rodents, and the presence of water on the site. Then, I find out what are the annual temperatures and changing climatic conditions of the site? After I have collected everything there is to know about the site and the owners' aspirations, I ask the most important question: What would nature do if it had the same restrictions and requirements that I have? What would nature create? And what would be the optimal design solution? The study of insects gives me insights into the mind of nature because insects are the great builders of nature . . . Insects, and other of nature's creatures, use only found materials of the site to create a habitable design that is comfortable and efficiently allows the insects to complete their daily tasks. Insects are not influenced by what they have done in the past or by fashion or trends; they fully understand their immediate surroundings and create accordingly. And this explains why my building designs do not follow fashion or style; they are the direct*

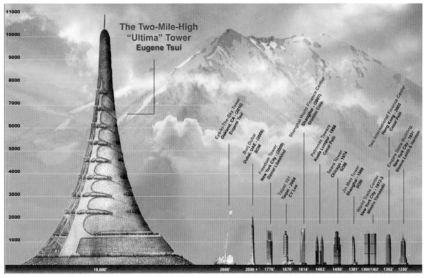

Insect-inspired architecture by Eugene Tssui:
Watsu Health and Medical Center (Middleton,
California; honeypot ants and oak galls), and the
two-mile-high Ultima Tower (termite mound)

result of the needs of the owners and the features of the site, the climate, and the intelligence of the insects and organisms that inhabit the site . . . Insects are the great communal habitat builders of the earth. They are our teachers. We must learn from them how to build without the need of electricity, highly intensive materials, and mechanical methods that pollute the earth.

Tssui and I talked about how people tend to adopt designs that are seen as safe or conventional, even if those same people are attracted to unique, ambitious, or novel works. Clients and investors may not leap to fund some of Tssui's grand-scale projects, but that does not stop him from drafting designs that dwarf anything ever manufactured by our species. Inspired by the giraffe-tall mounds built by African termites, Tssui has plans for a fifty-three-billion-cubic-foot, five hundred–story and two-mile-tall Ultima Tower. The design aspires to reduce the appalling sprawl of residential and commercial dwellings that is gradually replacing natural sites across the planet. More and more, I am convinced we will have to adopt Tssui's approach to learning from nature, and fundamentally rethink our architectural practices. Tssui is a force of irrepressible action, and just what the world needs.

Is it too outrageous to model our buildings after termite mounds? Already done! Shop in the temperature-controlled Eastgate Centre in Harare, Zimbabwe (opened in 1996), and you would benefit from architect Mick Pearce's attempt to imitate the physiology of *Macrotermes michaelseni* African termite mounds. Though the building succeeds in some respects, the architect emulates a now-outdated model of how a termite mound actually works. Instead of using a cooling system in which outside air rushes up through chimneys to ventilate the nest, biologist Scott Turner and engineer Rupert Soar have formulated a model that suggests these termite mounds act much more like a human lung. The mound's walls are porous interfaces through which stale air and fresh air are exchanged. Termites are the lung's alveoli, and the mound is an extension of the termites themselves. By relocating soil and respiring, the termites are part of a dynamic entity, operating on principles far different than those practiced by engineers and architects today. We could continue to learn from termites and construct walls not as barricades, but as part of a system that works in harmony with and responds to changing environmental conditions. Sustainable architecture is achievable, and insects have the answers.

Nowhere is the insect influence on human architecture more obvious than the hexagonal latticework of honeycombs. Many buildings, old and new, are modeled after hives humans have designed for honey bees (think of the classic, dome-shaped skeps—traditional beehives made of straw—that symbolize nests in cartoons), but wax honeycomb built by honey bees themselves offers incomparable stacking and packing potential. Those six-sided cells offer an efficient strength-to-weight ratio that lends itself well to walls, thermal insulation, and domes. Frank Lloyd Wright's Hanna-Honeycomb House (begun in 1937), now a National Historic Landmark, was built on a floor plan of hexagonal modules. The Climatron greenhouse at the Missouri Botanical Garden is a dome composed entirely of hexagons. Formstelle (Format Elf Architekten, 2013) is an office building in Töging am Inn, Germany, that receives just the right amount of daylight, thanks to a perforated pattern of hexagons that make up the aluminum facade. Honeycombs grace buildings from Izola, Slovenia (Honeycomb Apartments, OFIS Architects, 2005), to Singapore (DUO, Büro Ole Scheeren, 2018), and Belo Horizonte, Brazil (Josefine/Roxy Nightclub, Fred Mafra, 2011), to Surat, India (Hive, OpenIdeas Architects, 2019).

Is there an opportunity for us to build collaboratively with insects? Do hints of hybrid human-insect architecture exist? Kazuo Kadonaga (remember those 110,000 cocoons filling a cityscape of grids?) has employed thousands of silkworms to help build several large-scale silk pieces that hang like living, dynamic banners. Neri Oxman's group at the MIT Media Lab attempted something similar by having silkworms spin silk over geometrically arranged human armature to form silk "pavilions" (2013 and 2020). There is also the option of building insect robots that then build insect-style architecture at human scale. Justin Werfel and colleagues at the Wyss Institute for Biologically Inspired Engineering have imagined building robots

that act somewhat like autonomous termites, collectively constructing buildings for humans.

We can build something to resemble an insect's body, as designers of the Tiszavirág Bridge in Hungary did with arches and cables alluding to wings, and truss to the segmented body of the *Palingenia longicauda* mayfly. A more explicit example is Grasshopper's Dream, a café in South Korea with two stacked train cars painted green, and decorated with giant legs, antennae, and eyes. We can also learn from an insect's body and design something functionally relevant, like a building in the desert that passively collects water, modeled after a fog-basking darkling beetle (Las Palmas Water Theatre Building's conceptual design). And then there is the wealth of lessons we can learn from architecture created by insects, like the strength, kinetics, space efficiency, or thermal properties of mounds, weavings, and combs. As with all areas of biomimicry, these are early days for emulating insect architecture.

BECOMING THEM:

METAMORPHOSIS

JITTERBUG

Day-in-day out you rue
blue self-doubt accruing
all about you. You tire of the dire
cost the bugaboo of being bossed
requires of you. You've lost your fire,
desire, your high wire daring-do.
How now to clamber off the mat,
find your zing, wind the mojo back
into your mainspring,
get your lines uncrossed,
your linguini sauced?
Rather than be a quitter,
a sideline sitter, petrified
in amber limbo, do
what crawly creatures do
when suffering the humbugs
same as you.
Intercommunicate.

Transubstantiate.
To jump-start your ticker
tune into the quicker rubato
of some other's upbeat vibrato.

Cast off the dim glow, the grim
woe, let it enthuse you, infuse
your circumstance anew.
Get on your feet. Go!
Dance!
(Danse if you're in France—
à la Rimbaud perchance?) Dance
like a bee's glee at a rose's
sweet sensation, like a moth
drawn to an irresistible temptation.
Whirl, romp, hop,
bop, stomp, twirl! Forget
you've ever had a worry in the world.
Dance, limbs akimbo, feet
a-flurry, as if the floor is aflame,
as if it's your last chance, hurry.
Cut that rug.
Put your all into it,
dance
like a jitterbug!

—BILL HARRIS,
11 FEBRUARY 2023

Four legs good, two legs better.
—GEORGE ORWELL, *ANIMAL FARM*

Six legs best?
—ANDREW SUAREZ, REVIEWING *SIX LEGS BETTER:*
A CULTURAL HISTORY OF MYRMECOLOGY

Our bodies are soft, we have no wings, and we suffer from an embarrassingly small number of limbs. Our skeletons are on the inside, and our machinery for sensing the world shares little ancestral commonality or physical likeness with an insect's. How could we ever hope to competently imitate one? Pause. You might be wondering not how, but *why* anyone would ever consider flirting with such an inherently inhuman pursuit. Pretending to be an insect seems silly, or perversely antihuman. Yet, reasons for imitating an insect are many, and need not entail the rejection of our humanity. An insect has the ability to shock, disgust, bewilder, transfix, or amuse, so to imitate an insect offers the potential to evoke strong emotions in others, or in ourselves. Insects have been solving problems far longer than we have, and can teach us how to minimize traffic, manage our economy, and make decisions as a collective. For some, the act of imitating insects serves as a means of paying respect to nature, or finding greater intimacy with our surroundings. Imitation could be our closest avenue to channeling an insect's essence, portraying its sentience, or conveying its ecological importance. To know an insect, one must become an insect.

It will never be possible to truly comprehend what it means to be an insect, just as the philosopher Thomas Nagel concluded that we can never know what it subjectively means to be like a bat, but we can try. Anatomical disparity aside, we have factors working in our favor. As a species, we have

Compound Eyes by
Lea Bradovich

encountered insects for as long as we have existed, and gained an intimate familiarity with some insects' life cycles and behaviors. Insects take so many shapes and forms, and move in such varied ways, that we can also be choosy when selecting what to imitate.

> *Imitation is the sincerest form of flattery*
> *that mediocrity can pay to greatness.*
> —OSCAR WILDE

If Wilde was right with his twist on an old proverb, then our attempts to imitate insects could be perceived as our paying tribute to insects' superior splendor. Our motivation to imitate, or pay tribute, can be personal, emotional, intellectual, spiritual, or purely utilitarian. We cross the line that separates us from our distant relatives in the way we dance, act, sound, dress, and even in the way we fight.

FIGHT
LIKE AN INSECT

I'm going to float like a butterfly and sting like a bee.
His hands can't hit what his eyes can't see.
—MUHAMMAD ALI

A praying mantis is perched on wiry stilts for legs. Her body remains motionless as her head swivels to face movement on a nearby branch. A massive cicada, unaware of the predatory threat, betrayed his camouflage with a readjustment of a leg. Several thousand lenses make out the cicada's form as the praying mantis lurches slowly and silently forward, bobbing back and forth like a leaf in a slight breeze. When in range, anticipation builds as she prepares to launch her greatest predatory asset. Raptorial front limbs, lined with spines, spurs, and tubercles, unfold, extending in an instant to grasp the cicada, crushing and distorting parts of the bug's body. The cicada erupts with a shrill noise, a desperate departure from his usual mating call, as he flaps his wings to extract himself from the mantid's muscle-bound vises. The mantis bends her long prothorax and head forward without pause, and with "grotesque calm," as Frank Lutz described it in his book *A Lot of Insects*, begins the process of chewing through exoskeleton and muscle.

The brutal, gradual reduction of the cicada continues as the rattling wings and shrieking song attract the curiosity of an onlooker. Sometimes, the noise or flash of captured prey can attract other predators, and

Stills from mantis-style kung fu animations by Shon Kim

became the most distinctive characteristic of the "Northern Praying Mantis Style".

Techniques of the Praying Mantis Style — created from an inspiration after observing movements of the forelegs of the mantis and steps of the monkey.

LEE KAM WING MARTIAL ART SPORTS ASSOCIATION

4 반란수(搬攔手)

This comes out faster when you can push like this
이렇게 밀 때 빠르게 나갑니다

I can shake any part of my body very easy and fast
신체 어느 부위라도 이렇게 쉽고 빠르게 흔들 수 있습니다

a scuffle between competing predators can allow for escape, but not here. The insects' audience is Wang Lang, a monk and a fugitive in dire need of inspiration and change. Wang Lang had suffered a series of defeats sparring a fellow monk named Feng, and was taking a break to study Buddhism. Only a new approach to fighting, a new way of using his body, could improve his chances against future adversaries. As a spectator to the grisly insect conflict, it would be easy for him to relate to the cicada—helpless, incapacitated, and vanquished, but dwelling on the prey's predicament would not be a recipe for securing future success. Instead, by focusing on the tactics of the mantis, he imagined a path to his own victory.

It is roughly 1620 and the Ming Dynasty is being forcibly replaced with the Qing Dynasty by the Manchu clan. Wang Lang was planning to incite a rebellion with monks and disciples of the Shaolin Temple, but plans were quashed by the new empire and the temple was burned to the ground. Legend tells of Wang Lang and other survivors fleeing and, eventually, settling in the mountains of Shandong province. It is here that he practiced his martial skills with Feng and repeatedly experienced the shame of defeat at Feng's hands. It is here that he became rapt by the unfolding insect drama.

The allure of watching insects battle was nothing new. Fighting matches between insect combatants had been staged since at least the Tang Dynasty, and even today annual championships and underground gambling parlors pit male crickets fed special diets and separated by weight class against each other. Mantis fighting had been part of this tradition. Even Bruce Lee's character in the martial arts cinematic classic *Enter the Dragon* wins a match by betting on the smaller of a pair of fighting mantids. But could a solitary, predatory insect most closely related to cockroaches reveal movements and abilities transferable to humans for martial application? Many other animals have inspired martial art styles, including the tiger, leopard, boar, dog, monkey, snake, eagle, and crane, but no serious attention had ever been paid to insects.

Wang Lang took the mantis home and continued learning from the swift and nimble insect, convinced he

Mantis kung fu, as practiced by Bruce Lee. Other martial arts film stars, like Jackie Chan and Jet Li, have also trained in mantis style.

could adopt mantis postures and rapid striking motions in his fighting. He prodded the mantis with a piece of straw, watching as the mantis pivoted with an evasive body slant, or lunged with her grasping forelimb. He likely saw in the mantis what looked like patience, confidence, and strength that could serve a martial artist well.

Our arms are not naturally armored or spiked, but they are free to fold and extend, to lunge, parry, strike, hook, or grasp in ways that can mimic a mantid's. But what of a mantis's other legs? They are little more than static props for the body. Wang Lang overcame this weakness by observing and incorporating monkey movements, animating the legs with hops, low kicks, and sweeps. He fused mantis with monkey, often employing simultaneous mantis arms with monkey steps, one feigning while the other strikes. Wang Lang incorporated other styles into this fusion in what blossomed into a new, "complete" martial art. Northern Praying Mantis later branched into a variety of styles, distinguished by soft and hard approaches to combatting opponents' advances. Southern Praying Mantis, a distinct Chinese martial art arising from comparably uncertain origins and legends, spawned its own branches of study. Mimicking a mantis for martial gain is now practiced in schools around the world, as well as in virtual worlds. With physically less at stake, you can fight like a mantis by assuming the identity of the video-game characters Gen (*Street Fighter*), Lion Rafale (*Virtua Fighter 2*), Larcen Tyler (*Eternal Champions*), or the *Mortal Kombat* fighters Shujinko, Shang Tsung, and teleporting, vortex-projecting, and hat-hurling Kung Lao.

And, yes, when Feng next sparred with Wang Lang, he was soundly defeated by the man who had so carefully mastered the ways of a mantis.

DANCE

LIKE AN INSECT

A transformation is taking place in the life of a girl. As a member of the Ju|'hoansi (Zu|'hoasi) community in the western Kalahari Desert of Africa, she is participating in a ceremony that will alter her life. She is about to experience her first menstrual period, and for her this means joining women in her San group to perform rituals in a "cleansing" hut. What she experiences in this hut is a rite of passage that signals maturation and a change of status in her community. Fittingly, before entering the hut, she first imitates the motions of an immature insect approaching metamorphosis. Importantly, she does not do this alone. She is joined by fellow San women who coordinate their movements to collectively dance as a single larval insect—a caterpillar. The Ju|'hoansi women form a line that bobs and undulates, taking incremental, rhythmic steps to the hut. This subtle, beautiful dance requires moving in synchrony, in harmony, to the girl's symbolic equivalent of a cocoon.

That is, if we trust the description accompanying a single eighteen-second video uploaded by "elynngo" to YouTube in 2007. Even Arnold van Huis, who cited this video in an article about the "Cultural significance of Lepidoptera in sub-Saharan Africa" cannot confirm details about the dance, nor can anthropologist and founder of the Kalahari Peoples Fund, Megan Biesele. Biesele, who has studied Ju|'hoansi practices for decades, is cautious in her assessment: "It is possible that there is a reference to girls becoming women in this dance, but there is no definite evidence for this interpretation in the anthropological record." Why, then, do I begin this chapter with a dance mired in questions? The lack of certainty invites the curious mind to wonder what else is out there, hidden from the rest of the world. If the Ju|'hoansi found inspiration in a caterpillar and the dance has eluded the anthropological record, we can never know how many other dances owe their origins to insects. Dances could be confined to small communities, be short-lived, or symbolic and less obvious to outsiders. The original intent or source of inspiration for a dance could have been lost across generations. Though no resource or organization can tell us how many insect behaviors reside cryptically within traditional dances, this will not stop us from exploring some of the more spectacular, culturally enriching, and well-documented examples of insect dance.

Before we take this tour of vertebrates performing as invertebrates, however, let's first distinguish between dancing like a bug and dancing as if bugged by a bug. In Samoa, an invasion of mosquitoes gave rise to a traditional dance called the fa'ataupati, or Samoan slap dance, in which men forcefully slap their bodies as if in response to biting mosquitoes. The slaps, claps, and stomps rhythmically replace any need for instruments. Italians traditionally dance the tarantella, historically a symbolic response to the bite of the tarantula wolf spider (*Lycosa tarantula*). Though the wolf spider's bite is not dangerous, it was believed to cause tarantism, a hysterical state that required frenzied dancing to remedy the condition. George Balanchine adopted the folk dance's name for a high-stamina ballet piece, performed to tarantella music. Likewise, the bug was a 1960s fad in which dancers imagined a bug crawling on them. One dancer would pretend to throw a bug, eliciting dancing antics by the recipient of the imagined creature. The process would repeat, resulting in a succession of dancers contorting themselves in response to the fictive crawler. This is different than the jitterbug, inspiration for Bill Harris's poem at the opening of this section. Jitterbugging at a juke joint can look like "the frenzy of jittering bugs," according to dancer-singer-songwriter Cab Calloway. Dancing *like* a bug means striking insect-like postures or behaviors, and such expressions are far from rare in the world of dance. They can be spotted on dance floors and theater stages, and in village squares, royal courts, pavilions, and kivas alike.

Hopi figure (kachina
doll) of the Butterfly
Maiden, Polik-mana

Imitating an insect is a staple for some traditional dances. Take a journey to a nation within a nation within a nation (Hopi Nation, Navajo Nation, United States), and during an annual two-day ceremony, the Hopi celebrate nature, renewal, and the coming harvest by dancing the butterfly dance. Unwed girls are the primary participants, and one of the dancers represents Polik-mana, the Butterfly Maiden spirit. As is customary on the Hopi Reservation, no photographs can be taken or sketches made of Polik-mana, but kachina dolls and pottery depict her, and a mountain on Venus bears her name. A relatively recent (roughly century-old), intertribal social dance, the fancy shawl dance, is also commonly called the butterfly dance. Women keep at least one arm (wing) extended as they remain light on their feet. The dancers wear elaborate shawls with elements of their apparel often fringed, flared, or beaded.

Butterflies also appear in the wolf dance of the Kwakwaka'wakw of the Pacific Northwest coast. Women imitate butterflies, dancing with bows for antennae, during a potlatch as they dispense gifts to other women. Kwakwaka'wakw ceremonies can feature dancing mosquitoes, children mimicking the moth that brings sleep, or bumblebees. Marianne Nicolson, artist and member of the Kwakwaka'wakw Nation, described the bumblebee dance performed by young children in association with her art piece *Waxemedlagin xusbandayu'* (*Even Though I Am the Last, I Still Count*, 2000):

Amongst the Musgamagw Dzawada'enuxw, it is often one of the first dances a child participates in during the Winter Ceremonial. A father and mother bee lead progressively smaller bees out onto the dance floor one by one. When the children are led back into their "beehive" at the end of the dance, one child is discovered to be missing. The father bee circles the floor four times searching for this lost child. On the fourth round the child is found hidden amongst the spectators and is led home.

The influence of insects on dance is certainly not limited to North America and the mysterious caterpillar dance of the Ju|'hoansi. The Bobo people of

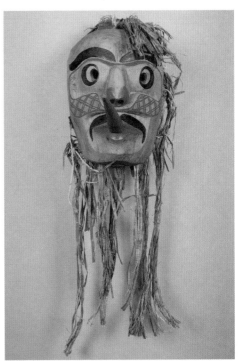

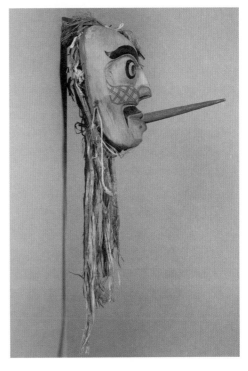

Kwakwaka'wakw ceremonial bumblebee mask made by Joe Seaweed, and mosquito mask (artist unknown)

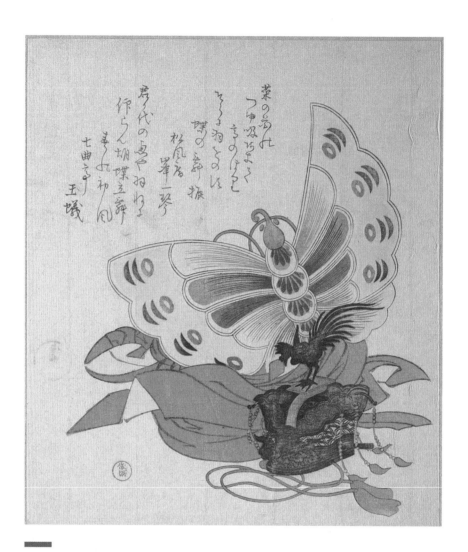

菜
の
花
の
つ
ゐ
双
ち
ま
く

そ
こ
ら
に
は
蝶
の
舞
振

松
風
客
峯
一
琴

岩
代
の
鬼
や
ね
け

作
ら
ん
胡
蝶
立
舞

七
曲
さ
す
玉
璣

Costume for the Butterfly Dance (Kochō no mai) from the Spring Rain Collection (Harusame shū), vol. 2, created by Kubo Shunman (1810s, woodblock print)

West Africa's Upper Volta dance to bring rain to ensure a successful harvest, and it is a butterfly dancer, with a massive winged mask, who performs this important dance. The mask whirls like a propeller to activate a rain charm, and dancers "tickle the clouds with their tall masks, to make the sky weep with laughter." In Japan, the kocho (butterfly) dance was performed in the imperial court, and the Ainu of Hokkaido imitate grasshoppers in pattaki upopo. A Filipino folk dance called the alitaptap is said to mimic lightning beetles.

Dancers in Cambodia perform as jewel beetles in robam kamphem, and dance as praying mantids in robam kandob ses (praying mantis dance). As mantids, the dancers strike praying mantis poses while wearing mantis colors, knocking coconut shells in the process. In Papua New Guinea, raptorial forelegs and body extend from the head of a Sulka dancer as he performs a differ-

Hemlaut mask by a Sulka artist from East New Britain Province, Papua New Guinea, features a praying mantis sculpted from spongy plant pith (circa 1910).

ent praying mantis dance. The dancer jumps up and down in place, and tilts to display the mantis mask's umbrella, adorned with painted motifs.

Australia is distinguished by having a wealth of danceable insect biodiversity, with at least ten insect totems and their respective ceremonies. Each totem is traditionally honored to increase their numbers, and rituals can invoke insect imitation. For the witchetty grub ceremony, participants sing about the different developmental stages of the insect (typically the moth *Endoxyla leucomochla*), from immature to adult. Like the "Maegwa" emerging from a cocoon, they wriggle their way out of, and then back into, the wurley,

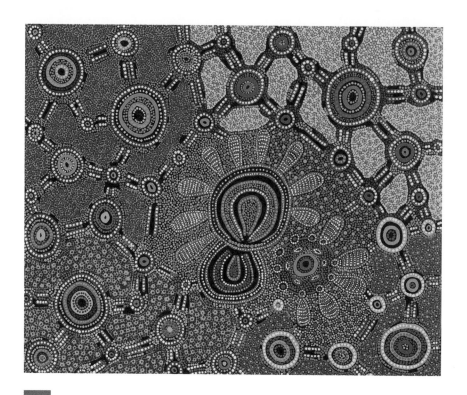

Witchetty Grub
Dreaming by
Jennifer Napaljarri
Lewis. Napaljarri
and Nungarrayi
women are col-
lecting "ngarlkirdi"
(witchetty grubs) in
Kunajarrayi (Mount
Nicker), Australia.

or hut. The honey ant (*Melophorus* spp.) ceremony, by contrast, features a performer decorated with representations of ant tunnels traveling from neck and arms to nest chambers on the body.

Even with death, insects dance. In a typical ceremony following the death of a fellow Aboriginal Australian, caterpillar dancers are separated from snake dancers by a fire as the two prepare for combat. Iguana dancers put out the fire, signaling a truce, which pays respect to the recently deceased. It is worth noting that the witchetty grub and honey ant are not only spiritually important for Aboriginal Australians, but they are also traditional sources of sustenance. Witchetty grubs are a valuable source of protein, and honey ants are a sweet treat, laboriously excavated from the earth.

Modern and contemporary dance continue to borrow from insects, often with choreographers carefully examining, then interpreting insects' movements. Doris Humphrey choreographed the dramatic birth and death of a

ABOVE: Femme fatale insect dancers from *Gossamer Gallants* (Paul Taylor Dance Company, 2013)

Minako Seki was *Insect* (2013) for a solo performance fusing butoh with contemporary dance.

queen bee, following lessons from Maurice Maeterlinck's book *The Life of the Bee* in her 1929 ballet adaptation. Akaji Maro brought an ant into his butoh studio for his troupe to observe and mimic, which birthed *Mushi no Hoshi—Space Insect* (2014). Merce Cunningham studied insects under microscopes as part of his exploration of movement, and Paul Taylor claimed there was little difference between creating dances and arranging collages of insects. Taylor's dance arrangements include *Gossamer Gallants* (2011), featuring winged, antennae-bearing female dancers pursued by their male insect counterparts. With a nod to femme fatales in nature, the females turn against the males, leaving all but one escapee in a heap of bodies on the stage.

Sixty years prior, Jerome Robbins choreographed *The Cage* (1951), a ballet in which female insects prey on their males. Eliciting awe and some discomfort, Nora Kaye seemed to shed her human husk in exchange for embodying an insect when she performed as lead dancer. Robbins described it this way:

> She didn't ever play human or have human responses. She was much more terrifying & unearthly. She performed the role quietly. With a beetle's eyes & no expression. As one cannot read into [the] eyes or thinking [of] an insect she remained appalling in her surrenders, instincts, and actions— an extraordinary creature—not a ballerina doing ketchy movements.

Loïe Fuller, an experimenter and pioneer of modern dance who introduced colored gels to stage lighting, performed *Papillon* (*Butterfly*) in 1892. Always inventing new, visually exhilarating ways of expressing herself, Fuller forged a friendship with Marie Curie, the only person to be awarded Nobel Prizes in two different fields (physics and chemistry), initially with the hope of acquiring radium to create a radioactive dress (Curie advised against it). For *Papillon*, Fuller fabricated her own silk costume with bamboo staffs sewn in to create billowing, wing-like extensions that caused a sensation in France, and inspired Jody Sperling's and Elizabeth Aldrich's *The Butterfly Dance* (1997). The list of insect dance hits goes on. Hybrid insects become gradually disfigured in anatomically glorious fashion in Cecilia Bengolea's

Portrait of Loïe Fuller
as a Butterfly *in 1901,*
gelatin silver print

and Florentina Holzinger's *Insect Train* (2018), a dancer twitches and rubs legs together while on trapeze in Gao Yanjinzi's *Insect* (*Oath-Midnight Rain*, 2006), and Julia Oldham hops and flutters while barefoot in the woods in *The Timber* (2009). In South African ballet, dancer and choreographer Frank Staff based *Mantis Moon* (1970) on the Mantis god of the San people—the same god that created night by piercing the gallbladder of a slain antelope (after not heeding the organ's warning that the act would engulf him).

There's a butterfly stretch, a butterfly stroke, and, the astute reader will have noticed, a preponderance of butterfly dances. As Jamaican scholar Sonjah Stanley Niaah recounts, a once extremely popular reggae dance takes the form of a butterfly:

> *The butterfly is danced with bent knees, a characteristic feature of African and diasporic movement patterns, with the feet flat to support the dynamic displacement of the hips, shoulder girdle, and legs. The knees, which open and close fluidly on a horizontal axis, mimic the flapping of the butterfly's wings in flight.*

Ballet dancers flit across the stage as butterflies in George Balanchine's *A Midsummer Night's Dream* (1962 premiere), and join to form an undulating caterpillar in Christopher Wheeldon's *Alice's Adventures in Wonderland* (2011 premiere). Where would breakdancing be had no one dared send their body in a rippling motion across the floor to first perform the classic "caterpillar" (aka worm, centipede, or dolphin)?

Of all the arthropod dances, one stands out to me as the acme of arthropod imitation. Humans typically have only four limbs, yet Milena Sidorova, choreographer and soloist at the Dutch National Ballet, was able to contort her body, stretch, wave, and bob in a manner that, on a continuum between eight-legged spider and human, was far, far from human. Sidorova first imitated a spider in her living room at thirteen years old, and performed as an arachnid soon after (*The Spider*, 2000). She is frequently asked about mistakes commonly made when attempting this dance. Her website response: "If the opening pose isn't done properly, you're more likely to

look like a frog than a spider. Your knees need to be aligned with the hips and they should be higher than the back. This takes both flexibility and exercise." You might find this to be an understatement after watching the eighty-one-second video of her arachnid feat. Compelled by her spider-like abilities, computer-graphics animator Robyn Luckham invited her to don a motion-capture suit to model for Baron Vladimir Harkonnen's pet humanoid spider in the blockbuster film *Dune* (2021).

Milena Sidorova, performing *The Spider*

From the Ju|'hoansi traditional, tandem movements as a caterpillar to Milena Sidorova's solo spider stage performance, human dancers have remodeled their own behavior to more closely match what they observe in arthropods. And for good reason. Arthropods move in ways that can surprise, mesmerize, tantalize, or terrorize. Arthropods also dance.

Honey bees perform their famous waggle dance to communicate to nestmates how to find food, water, tree resin, or a new home. Dancers feverishly waggle their bodies from side to side as they make their way across the vertical honeycomb. Dancers circle back, repeat, circle back in the opposite direction, and repeat. The dance continues as a dramatic, whirling figure-eight. The direction they waggle is not random. The angle of the "waggle run" relative to the vertical direction is identical to the angle between the site they are advertising and position of the sun's azimuth (direction as if the sun were on the ground and not in the sky). Because direction is usually not enough for a naive follower to locate a destination, the dance also encodes distance information. The duration of each waggle run passes along this information. A bee following a waggle run lasting three seconds will have to fly much farther than if following a waggle run lasting one second. The waggle dance is ritualized and predictable, but the choreography is not fixed. A dance's direction depends on the angle of the sun, its precision depends on what she is advertising and how far away it is, and the intensity of her dance depends on the quality of the commodity being advertised. Each bee is also unique, and displays personality that can be studied and documented, including as it relates to her dance.

For other arthropods, dancing means courtship. Scorpions perform a promenade à deux, featuring a male lead, with pedipalps (pincers) clamped on pedipalps. The dance involves guiding the female over a sperm packet he has left on the ground, and has been known to include cheliceral (pinching mouthparts) kissing, and even some stinging. Jumping spiders also dance during courtship, and theirs is a mix of generating vibrations through the substrate as they wave tufted limbs, lift their bodies, lean and shimmy to one side, then skitter to the other. Their bodies sometimes iridesce, or pulse to a drumming beat beyond our range of hearing. Arthropods wiggle, scurry, flitter, dart, and scramble. The flight dynamics and locomotory patterns vary tremendously, so I will take this opportunity to advocate for more (bio-) diversity on the dance floor. Observe a different insect's or arachnid's movements, interpret them, and express yourself through arthropod mimicry. If imitating a butterfly or caterpillar can set the world on fire, think what imitating a stalk-eyed fly or (the aptly named) dance fly could do.

ACT
LIKE AN INSECT

Go to the ant, thou sluggard; consider her ways, and be wise:
Which having no guide, overseer, or ruler, Provideth her meat in
the summer, and gathereth her food in the harvest.
—*THE HOLY BIBLE*, (PROVERBS 6:6-8),
KING JAMES VERSION (1611)

As an entomologist who has spent most of his career on hands and knees
studying the highly organized and super-efficient colonies of ants
I am often asked the following question: What can human beings learn
from these insects? And the answer I give, with my fingers steepled,
*my mouth pursed, and my eyes squinted, is—**Nothing**. Not a thing and*
it would be a major Aesopian error to believe otherwise.
—EDWARD O. WILSON (1998)

Tens of thousands of honey bees have clustered on a tree branch. Exposed to the elements, their lives depend on finding a new home without delay. Scouts fly off to independently seek tree cavities and other possible homesites within several miles, and return with a dance, inviting others to assess their discovery. Each site means a different dance, and the surface of the swarm can be whirring with a variety of dances. The bees have days, possibly hours, to come to some consensus as to which dance advertises the best home. Tom Seeley, honey bee scientist extraordinaire, observes as the mass of bees gradually disassembles, scatters, and fills the air around him. The erratic swirling and buzzing cloud seems like nothing

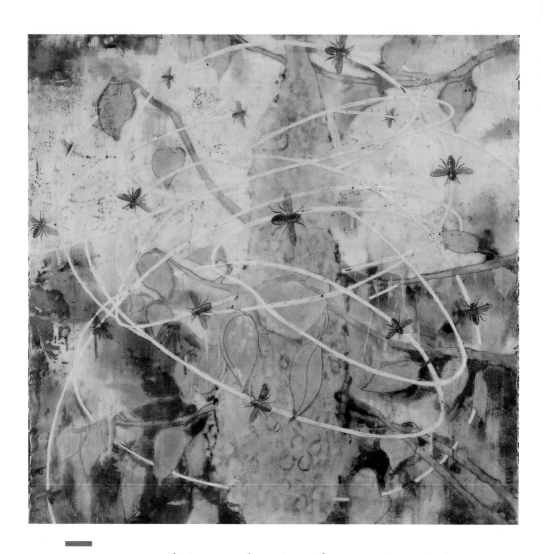

but pure pandemonium. After some minutes, Seeley can see through the chaos, however, and looks for patterns of activity that culminate in a unanimous decision by the amorphous clutter. A subset of worker bees streak above the others, rapidly flying in one direction. They slowly circle back, then repeat the high-flying, streaking behavior. Within minutes, the bedlam of individual bees becomes a collective entity, flying in unison in the direction of a chosen home, though fewer than 5 percent of the bees have ever seen it.

So effectively has natural selection shaped bees' collective decision making that when Seeley graciously stepped into the role of chairperson in his department's faculty meetings at Cornell University, he attempted to apply lessons learned from the bees to his human colleagues. But in what possible way could humans benefit from acting like that swarm of bees? We can't fly, we don't live in hollow trees, and it isn't standard practice to dance when communicating directions to a destination. As Seeley reports in *Honeybee Democracy*, the following five lessons appeared to work quite well, at least in the context of making decisions in his department.

First, compose a group with shared interests and mutual respect, rather than of "clashing curmudgeons." Like honey bees flying to a single new homesite that benefits the entire swarm, human communities can work toward common interests, especially when reminded that common interests exist. Second, minimize the influence of a leader. If a human leader acts impartially, collective decisions—not unlike honey bees swarming in a single direction—can emerge. Third, seek diverse solutions to a problem. Honey bees dance for different possible homes, with no single dancer dominating the message. Seeley suggests that a sufficiently large group of humans with diverse backgrounds and perspectives who independently explore and then discuss a problem have the best chance of making effective decisions. Fourth, engage in an open competition of ideas, and build support among voters until a quorum (a minimum number necessary to conduct business) is reached. A swarm of honey bees reaches a quorum through honest advertising of homesites until a threshold of scout bees is recruited to—and then dances for—one best site. Finally, try to reach a quorum without wasting time. At some point, it is beneficial to come to a decision, and to do so accurately and cohesively can be challenging. Honey bees do this by making squeaking "piping" signals to the colony when a threshold number of responders to a single homesite has been reached, and Seeley achieved this through straw polls during departmental meetings.

The honey bees need to find a home they can agree on before a storm strikes, and that departmental meeting must reach a timely conclusion or productivity will lapse, morale will crumble, emotions will flair, and

Having set up an artificial swarm of honey bees, Tom Seeley is able to see if the bees will dance for and, ultimately, choose a box he has set up in the far-off tree to his right. Despite Seeley's efforts, that particular swarm made a different choice, flying in unison over a corn field to a site unknown to us (but possible to find, were we to decode the appropriate dance).

once-dignified scholars will turn into incoherent, savage beasts. Even when there is disagreement, we can join a consensus if we see it is in the community's best interests. We, as Seeley has observed, have opportunities to learn from bees, "because there are times where humans' interests *do* align."

Seeley's gentle experiment with colleagues was not the only case of his honey bee research guiding human activities. An algorithm, based on his understanding of how honey bee colonies optimally forage for nectar, was adapted and applied to the $50 billion web-hosting industry, according to the committee that conferred The Golden Goose Award to Seeley and four engineers from Georgia Institute of Technology. Although this application was unplanned and unexpected, others actively attempt to apply insect behavior to our own. Michael O'Malley began to see his honey bees as if they were running a successful business when he wrote *The Wisdom of Bees: What the Hive Can Teach Business about Leadership, Efficiency, and Growth*. Paul Ormerod found that economies often behave like living organisms and that we can learn from the erratic behavior of ants when he wrote *Butterfly Economics: A New General Theory of Social and Economic Behavior*.

Those erratic ants could get us out of a sticky traffic jam, seeing as they have fine-tuned their traffic for tens of millions of years longer than we have. Just observe leaf-cutter ants and you will see that, instead of bumping into each other all the time, they travel relatively smoothly to and from their nest. Those heading out to forage always make way for returning workers laden with leaf fragments, and though laden workers slow down unladen workers behind them, the unladen workers ultimately save time by going at the slower pace, rather than risking bumping into oncoming, outbound traffic. Not only that, it pays for them to carry less than a maximum load because the ants can deliver smaller loads faster, which prevents a bottleneck at the nest. Ant biologists writing about "traffic rules," often allude to human municipal planners, engineers, and how ant behavior could inform our behavior.

Ants can solve difficult optimization problems, so are worthy of study when it comes to formulating optimization algorithms. Eric Bonabeau, who once studied ants at the Santa Fe Institute, now considers the actions of individual ants and the collective outcomes of colonies when he applies "agent-based modeling" to design forecasting and optimization tools for businesses. Companies are beginning to adopt "swarm intelligence" rules, or "swarm logic," when designing energy grids, by applying decentralized controllers that communicate wirelessly and distribute energy as needed. It appears that the leaderless actions of multitudes might offer solutions to problems we face, if we are able to glean the appropriate messages.

That's it—we should all act like insects! So far, it seems embracing our inner ant would be nothing but beneficial. Tom Seeley's colleagues remained civil, and the algorithms paint a pretty picture of efficiency and productivity. *Aber, Achtung!* (Caution!) E. O. Wilson's warning, earlier, should give us pause. Indiscriminate adoption of insect behavior would spell disaster. Just watch ants kidnapping the offspring of other species to raise as their own workforce (independently evolved in several ant lineages) for a lesson in which behaviors to steer clear of. Depending on the behavior you wish to observe, and the lessons you wish to impart, insect stories offer conflicting messages and morals, advocating for cooperation or conflict, peace or war, feminism or sexism, equity or hierarchy, fidelity or cheating, altruism or selfishness, harmony or revolution, freedom or slavery. Which will it be?

In the thick of World War II, W. C. Allee published in the journal *Science* ten types of evidence supporting the biological basis for cooperation. Others sought similar messages in nature, and even when confronted with notoriously unfriendly ant-termite interactions, would focus on isolated instances in which ants and termites live together, and framed their relationship as one of mutualism, offering each other protection. If ants and termites can live together, then so can we, seemed to be the implication. Auguste-Henri Forel, ant biologist who studied both human and ant brains, had more pessimistic impressions of us. Forel insisted that humans are absolutely incapable of living as ants do, and that we invariably and tragically rely on chieftains rather than adopt ants' "admirably co-ordinated state of anarchy." Though he wrote much about (ant) warfare in his 1928 book *The Social World of the Ants Compared with That of Man*, he did prescribe a recipe for peace, which combined his adherence to socialism and his knowledge of social insects:

> *If you want peace, prepare for it by disarming all men and organizing them in accordance with a true labour socialism of peoples that elect for themselves a supernational authority, supervised by them and responsible to them. We should accomplish this by means of an international alliance based upon that of the polycalic formicaries and combined with that of*

the mixed colonies. Even then, we shall still have enough natural enemies outside the human race, and enough work within it to keep the national and supernational police quite busy.

Theater of Insects by Karen Anne Klein (2013)

Forel was not the only scientist to claim a biological basis for an ideal state of socialism or communism in the colonies of social insects. As Diane Rodgers reports in *Debugging the Link between Social Theory and Social Insects,* naturalist, illustrator, and educator Anna Botsford Comstock wrote with admiration about the "laws of socialism" followed by honey bees, as did biologist Thomas Huxley (called "Darwin's bulldog" for defending Charles Darwin's theory of evolution by means of natural selection). Forel, Comstock, and Huxley all saw a successful form of political harmony in social insect colonies, but Rodgers cautions that a person's social environment can influence their interpretation of natural processes. The fact that the honey bee "queen" used to be identified as the colony's king, and for centuries a monarchical model has been imposed upon honey bee colonies, should tell us a lot about how our thinking is grounded in our politics. Depending on the observer, an insect society can represent ideals attributed to a monarchy, to socialism, or to communism. How remarkable, then, that the very same species admired for exhibiting qualities ascribed to these political frameworks inspired the careful research and writing by Tom Seeley in *Honeybee Democracy*.

> *We're the masters of the world . . . The largest Ant State! . . .*
> *The largest Democracy!*
> —CHIEF ENGINEER (ANT), FROM ACT III, *THE INSECT*
> PLAY BY KAREL AND JOSEF ČAPEK, 1921

Admit it, when you turned to this chapter, "ACT like an insect," you weren't signing up for a political discourse or some abstract version of acting. You wanted human insects performing onstage. Well, there is no shortage of actors skittering like insects in theater or on film. We have already seen beautiful examples of humans acting like insects in traditional ceremonies and contemporary dances in the previous chapter. The lines between dance and theater are further blurred when we see the great dancer Mikhail Baryshnikov in one of the many adaptations of Franz Kafka's story *The Metamorphosis* (*Die Verwandlung*, 1915; performance in 1989), in which a beleaguered salesman "awoke one morning from uneasy dreams" and "found himself transformed in his bed into a gigantic insect." Baryshnikov trembled and agonized, scooted and scurried across a Broadway stage as the woeful human-turned-insect Gregor Samsa. Director Roman Polanski played the physically demanding role one year prior in Paris.

If Gregor Samsa transformed into a cockroach (there is much debate as to how to translate *"ungeheueres Ungeziefer"*), then he was not the only human-cockroach to appear onstage. Federico García Lorca introduced a cockroach that falls in love with a butterfly tragically grounded with a broken wing in *El maleficio de la mariposa* (*The Butterfly's Evil Spell*; 1920, Spain). The script was lost and recovered, but pages, including the finale, remain missing. Needless to say, the play does not end in the lovelorn cockroach's favor. Egyptian writer Tawfiq al-Hakim introduced an absurd cast of cockroaches in his play *Masir Sorsar* (*The Fate of a Cockroach*, 1966). A (cockroach) minister's son is killed by ants because he fell on his back (a sad fate befalling many a real insect). The minister proposes battling the army of ants, but his king questions the idea: "The ants know the discipline of forming themselves into columns, but we cockroaches don't know discipline . . . how many generations will the species of cockroaches be taught and trained to walk in columns?" Finally, Kafka's tale of transformation takes a different turn in Melita Rowston's *Cockroach* (2018), in which Greek myths (from Ovid's *Metamorphoses*), replete with gendered violence, are turned on their heads. As advertised: "When C wakes from a nightmare, she finds herself transformed into a monstrous vermin . . . Join C on a bloodcurdling bar crawl as this avenging exoskeletal anti-heroine rises from the ashes of abuse to become the true definition of a pest."

Cockroaches and fellow Ungeziefer join a much grander ensemble of insects sharing the theatrical spotlight. Butterflies, beetles, ants, crickets, moths, and an ichneumon wasp form the majority cast of Karel and Josef Čapek's satire *The Insect Play*. Ants star in Lee Breuer's and Bob Telson's musical *The Warrior Ant* (1986, in partial form), with bunraku puppeteers, Afro-Caribbean music, Middle Eastern dance, Moroccan dance, and West African storytelling. Mayflies are comically (tragically?) burdened with the knowledge that their courtship has a shelf life of one day in *Time Flies*, one segment of Dave Ives's off-Broadway hit *Mere Mortals* (1997).

If scientific accuracy (or insect sex) is the mark of great stage drama, then Isabella Rossellini's *Green Porno* series achieves the pinnacle of theatricality. In season one (2008–2010), Rossellini magnificently narrates and portrays the sex lives of six arthropods, including *Bee, Fly, Mantis, Dragonfly, Firefly,*

An ant queen–
human hybrid in
Anna Lindemann's
staging of *The
Colony* (2019)

and *Spider*, and channels the high drama at the core of each natural history tale. Her *Seduce Me* series follows suit, with titillating vignettes of *Bedbug* and *Spider*. "It's not a Las Vegas stage show, you know, where I arrive with a trunk of costumes," Rossellini has said of her acts, full of quirky staging, deadpan humor, and spot-on science. She may have been referring to bona fide Las Vegas acts, like Cirque du Soleil's *Luzia, A Waking Dream of Mexico* and *Ovo*, with insects as acrobats, contortionists, and trapeze artists.

Whether acting in a pared-down, scientifically accurate, solo performance, or a circus-style escapade, performers need training to convincingly embody insects. Rossellini researched animal behavior. The cast of *Luzia* watched "endless documentaries," and artistic director Marjon van Grunsven visited insectariums. To train his performers for a stage production of *Mount Olympus* (2015), the artist Jan Fabre employed the "insect exercise," in which actors would transform from predator (lizard) to prey (fly or spider), and slowly embody the arthropod in the process.

For human-as-insect performances on the silver screen, let's observe an amateur entomologist, Niki Jumpei, alone and hunting insects on a remote stretch of sand. Jumpei unearths an antlion larva, infamous for digging conical pits, within which ants fall to their doom, ensnared by the antlion's sickle-shaped jaws. Jumpei himself is later trapped by a woman who dwells in a pit, and is forced to bail sand as the dunes ceaselessly threaten to engulf them both (Hiroshi Teshigahara's *Woman in the Dunes*, adapted from Kōbō Abe's 1962 novel). Getting a hair less metaphorical, we have Seth Brundle, brilliant scientist who unwittingly fuses his genetic makeup with that of a fly as both he and the fly teleport from one telepod to a second in David Cronenberg's *The Fly* (1986). Brundle transforms into "Brundlefly," walking on walls, vomiting on food to digest it, and living his remaining hours more fly than human. While Brundlefly has delusions of becoming "the first insect politician," he, like his counterpart Andre Delambre in the 1958 original film, cannot control the raging fly within, and faces a grim end. See, if you dare, other human-arthropod transformations in *The Wasp Woman* (1959, 1995), *Invasion of the Bee Girls* (1973), *The Nest* (1988), *A Nightmare on Elm Street 4: The Dream Master* (1988), *Franz Kafka's It's a Wonderful Life* (1993), *The Acid House* (1998), *The Hole* (1998), *Earth vs. the Spider* (2001), *Mansquito* (2005), *Infestation* (2009), *Metamorphosis* (2012), *Bite* (2015), *Terra Formars* (2016), *Weresquito: Nazi Hunter* (2016), *Leprechaun 4: In Space* (2017), and *Royal Jelly* (2021), for starters.

If passively viewing others live life without vertebrae doesn't satisfy and you would prefer to have more agency as an arthropod, the gaming world may be for you. In *Dungeons & Dragons: Honor Among Thieves*, you, as the Druid Doric, can turn into a fly-size fly. In the life simulation game *SimAnt* (1991), you have the power to control one black ant at a time to drive out rival red ants (and pesky human homeowners). In *Bug Fables: The Everlasting Sapling* (2019), you can guide insects to seek ancient artifacts and attend the Golden Festival. If the mood strikes, you could roll a dung ball as dung beetle postal worker in *Yoku's Island Express* (2018), or swing along as a female peacock jumping spider to save your spider boyfriend from a satin bowerbird in *Webbed* (2021). Characters can be vaguely buggy (The Knight

in *Hollow Knight*, 2017), sacrifice anatomical accuracy for cuteness (Charmy Bee in *Sonic Heroes*, 2003), or surreptitiously suck blood from human inhabitants (as *Mister Mosquito*, 2001). If none of these roles appeals, you can always resort to playing the downtrodden Gregor, Kafka's pitiable protagonist in Ovid Works' *Metamorphosis* (2020).

DRESS
LIKE AN INSECT

Acting the part can help if you dress the part. Here, we will consider how to become an insect, one step at a time.

Step 1. Wear a piece of clothing that depicts insects. Butterfly, bee, and beetle clothing is everywhere. If you are more discriminating in your insect selection and wish to display a subject from a less familiar order, diversity is limited. You would be hard-pressed, for example, to find angel insects, twisted-winged parasites, or mantophasmatodeans on a shirt in your preferred color. Variety does exist, however, and magnificently in Jenjum Gadi's Insect collection (spring/summer 2014, India), which features large prints of ants, bees, scarab beetles, flies, and antlions. For butterfly prints and patterns, visit Vivienne Tam's 2010 line of dresses.

Alternative: Make your own insect clothing by either sewing insect-patterned fabric, or painting your own insects on fabric, either freehand or using stencils.

Step 2. Adorn yourself with an insect accoutrement, like a piece of jewelry or accessory that is made to resemble an insect. Look to Victorian period or Art Nouveau jewelry and the insects can be fabulously decorative and anatomically accurate at the same time. Alexander McQueen covered his subjects in hundreds of butterfly wings made of cut, dyed, and painted feathers (Ensemble, spring/summer 2011). Jessee Smith (Silverspot Studio) fashions tiger beetle belt buckles and lapel pins of mosquitoes engorged with ceramic blood.

Alternative: Wear dead insects (or insect parts). Flip back to see traditional jewelry incorporating beetle horns and bodies, ant legs, and encysted scale insects. Iridescent elytra of jewel beetles are sewn into shawls, and famously shimmer on a dress worn onstage in 1888, painted by John Singer Sargent (*Ellen Terry as Lady Macbeth*, 1889). Fly legs as eyelashes? Jessica Harrison did it to create *Flylashes* (2010).

Second alternative: Wear live insects. I do not recommend tethering insects to your body (for their sake), but wearing "living jewels" is practiced in India, Sri Lanka, and Mexico, and there are accounts of people wearing bioluminescing click beetles or fireflies to share their limelight.

Step 3. Take a more permanent plunge with tattoos. Gallons of ink have been injected subdermally to produce lasting imagery of insects on flesh.

Alternative: Moth eyebrows. Shave your eyebrows and paint broad, short, wing-like, or slender, antenna-like "moth eyebrows (蛾眉)" as was the height of eyebrow fashion during the Tang Dynasty in China.

You can wear, or you can become. Our next steps guide you through the process of superficial transformation into insect forms.

Step 4. Change your hair, and not with a beehive hairdo, but with a nod to insect anatomy. A poem by Lu Chao-lin (circa 650) refers to hair of "cicada style." Women in China wore two coils of hair that, according to D. Keith McE. Kevan, resembled protruding compound eyes.

Step 5. Don an exoskeleton for the more complete look. Humans dress like insects for dance, theater, and film, as we've seen. There are traditional ceremonies that demand posing as insects, Halloween and other occasions where dressing as an insect is standard fare, and high-fashion insect transformations. During Fastnacht, a festival that takes place in parts of Germany, Switzerland, and Austria, devils, jesters, witches, and hairy bears can be joined by a clan of beetle-people. Monarch-people parade through the streets of Mexico during the Day of the Dead procession.

In the fashion world, Charles James designed *Butterfly* (1955), an 18-pound (8-kg) dress with a wing-like bustle composed of 25 yards (22.9 m) of tulle. Thierry Mugler took insect fashion in a more explicit direction, adorning models with enlarged eyes and antennae in Les Insectes

Monarch in Moda (2018), part of Jane Kim's *Monarch Migrating Mural* series in Ogden, Utah, is now a shoe by Le Mondeur.

(spring/summer 1997). Manuel Albarran has also clad his subjects with antennae, but creates exoskeletal elements in his "Metal Couture." Finally, there is the goliath beetle. Sarah Burton, as creative director of the Alexander McQueen brand, designed dresses using patterns from butterfly wings, as well as beetles. The wearer of her goliath beetle dresses, however, is wearing the pattern of the entire body of the beetle. In one version, only a human head disrupts the illusion. "It's about hyper femininity," Burton told *The Telegraph* in 2018, "and hybridisation. Is she a woman? Is she an insect? It's a story of metamorphosis."

Dress fashioned
after a goliath
beetle, designed by
Sarah Burton for the
Alexander McQueen
brand, 2018

RIGHT: Goliath beetle
drawing by Karen
Anne Klein (2023)

The Worst Elvis Imitator

The first time I went to work as a giant insect could have been a disaster. Halloween was my first day on the job at the American Museum of Natural History, and I traveled by subway from the Bronx in my Human Fly costume, with the head made of microcrystalline wax and a plunger serving as haustellate (spongy) mouthparts. I carried a giant human swatter. No one made eye contact in the subway car, but I sat down beside a comic book reader and struck up a conversation en route to work. I waited for the museum to open, and when the moment arrived for me to make a first impression on my new coworkers, I strode confidently through the entrance as security guards stood motionless with eyes wide and mouths agape. Passing through the palace gates unhindered, I made my convoluted way to the Exhibition Department and set to work. Needless to say, I never removed my mask, and none of my office mates knew what the new hire looked like.

Recycling dung beetle
(Bug Hartsock, 2023)

Two more alter egos. The costume with chest plate is made from the products of silkworms, paper wasps, honey bees, cochineal bugs, and lac insects.

The following Halloween was a bit less tidy as I dressed as a recycling dung beetle. While rolling a giant ball of recycled items through the museum's exhibit halls, my vision and hearing were limited, but I did notice a small boy point and yell, "It's a dung beetle!" Validated, I continued to roll my ball of recycled materials through the Hall of North American Mammals, around a corner, and into a work space where we were molding leaves for the future Hall of Biodiversity. Along one wall were the remains of a bird diorama that had been emptied and retrofitted to serve as my supervisor's office—a shell with a temporary wall and door where he could conduct official business. It was the perfect place to abandon my dung ball. Only later did I learn that a security guard had followed me through the museum, trying to alert me that rolling dung balls through the halls was strictly prohibited. I also learned that I had left a trail of recycled scraps in my wake. I discreetly cleaned up after myself. That night, after work and on the streets of Manhattan, it became clear that a face-down dung beetle without his dung ball is difficult for humans to accurately identify. Called a cockroach, shrimp, rock, and Elvis, I knew that to be an effective insect would take more concerted effort in the future.

BELOW: Jennifer Angus, sampling illustrations of dapper human-insect hybrids by Grandville (Jean Ignace Isidore Gérard), included textiles and real insects in a room-size installation for *Hunters and Hunted* (2012).

ABOVE: Jingu, of the *Insect Empire— Imperial Warriors Series No. 1,* is technically an insect that has become more human-like, and is one of a growing number of characters by Dave Beatty (digitally modified AI creation using Midjourney, 2022) who have adopted the ways of the samurai in a world where humans no longer exist.

Transforming into an insect, at least superficially, can serve as a way to commune with our underappreciated neighbors. If successful, it can also educate, amuse, or inspire others to reconsider their personal relationship with insects.

But is this enough? It may be time for a complete makeover, a transformation deeper than achieved with tattoos, dresses, or costumes.

Step 6. Body modification. Metamorphosis into a hybrid human-insect means realizing the dream (or nightmare, depending on context and personal preference) of embracing the insect body plan, or finding a place where the insect form and the human form can merge.

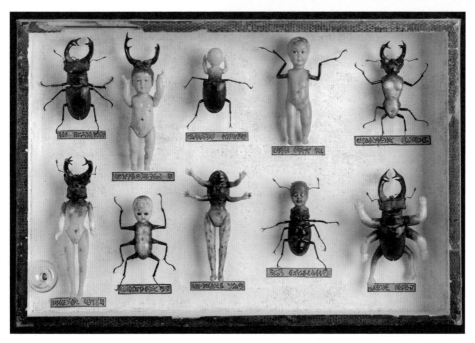

TOP: *Chalcosoma atlas* (2015) and *Cerogenes auricoma* (2000) by Laurent Seroussi

BOTTOM: *Manipulations* (2013) by Laurent Gauthier

SOUND
LIKE AN INSECT

One hundred sixty-five million years before Bach or the Beatles, a male katydid serenaded females in a forest of conifers and giant ferns. The frequency, at 6.4 kilohertz, was audible to many vertebrate ears. It was the Jurassic, so dinosaurs scampered through the same cluttered leaf litter from which this katydid rubbed his wings together to create his song. On each front wing was a vein with a row of ridges, or teeth, as well as a scraper—a pick, like what would be used to pluck very tiny guitar strings. He could choose which wing's teeth rubbed across the other wing's scraper, and switch as desired. His song burst forth as a series of pulses, each pulse (one strum of the scraper across the row of ridges) lasting sixteen-thousandths of a second. This katydid, *Archaboilus musicus*, sang at night in a pure tone at a single frequency to secure a private channel in what was undoubtedly an environment already crowded with different singing species.

How could we know anything about the song of an insect whose family went extinct at the beginning of the Cretaceous period? A beautifully preserved pair of fossilized wings was discovered in Inner Mongolia, with 107 visible teeth on the left wing and 96 on the right wing. By measuring the spacing of the teeth along the vein and comparing this with the katydid's closest living relatives, Jun-Jie Gu and colleagues could postulate about their habitat and their biology, and recreate a song not heard since *Protoceratops* gnawed on cycads. To hear this fossilized song played today is simultaneously exhilarating and haunting.

Insect sounds, according to David Rothenberg, philosopher and musician who wrote *Bug Music: How Insects Gave us Rhythm and Noise*, "may

How can a painter depict the chirps, trills, and sonic, reverberating chorus of insects on canvas? Charles Ephraim Burchfield portrayed insect songs in a language of visual waves breaking through the air in works like *Gateway to September* (1945–1956).

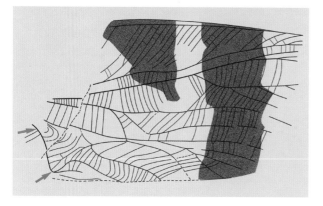

This 165-million-year-old fossil wing of a male katydid (*Archaboilus musicus*) was enough for Jun-Jie Gu and colleagues to resurrect the katydid's long-extinct song. Beside the fossil wing (*top*: ventral view / negative imprint) is a closeup of the file used to create the sound (*right*: dorsal view / positive imprint), and a drawing of the wing to show the location of the file (bottom: *between red arrows*).

be the very source of our interest in rhythm, the beat, the regular thrum." Singing insects avoid competition and confusion by singing different songs at different times, and the beauty of being immersed in a soundscape is that it changes as day turns into night, and the musicians can be singular songsters or collective, throbbing hordes. Like swarm intelligence lessons from honey bees in search of a new home, insect choruses and synchronized calls have no conductor or leader, so for crickets that sing in synchrony, they only do so because each cricket is behaving in response to the call of another cricket, and that one to the next . . .

The chirping, humming, whining, buzzing, clicking, thumping, and shrieking of insects are part of most peoples' lifelong experience with sound, so not only do insects end up in our music, on album covers, and in music videos, but we imitate them, either by manipulating our voices, or by playing our instruments in an insectile way. I wish this book were like a greeting card embedded with a sound chip so insect imitations would play as you opened each page. Instead, I will give examples of music that imitates insect sounds, and hope you are enticed to explore them in all their sonic glory. My selection will be unintentionally biased by my own experiences and search limitations, so I urge you to go beyond these and explore your musical preferences for imitative insect gems. We will begin with ambient allusions to insects, move on to brief imitations of insects, and close with full-on human-insect acoustic transformations.

Ambient Allusions

As with the use of insect products in art, medicine, and everyday life, the influence of insect sounds on human music could never be quantified. Cryptic references and ambient allusions to insects permeate music across its genres, political or geographic boundaries, and time. Growing up with insect sounds all around can unconsciously influence what a musician creates. On the other end of the spectrum, musicians explicitly look to insects and make their intent clear, including in a title. François Schubert's "L'Abeille" ("The Bee," 1860), Nikolai Rimsky-Korsakov's "Flight of the Bumblebee" (1900), and Béla Bartók's "From the Diary of a Fly" (*Mikrokosmos*, 1926–1939) all flit about not to mimic sounds produced by the flying insects, but in an attempt to put sound to the visually erratic flight of each.

Hints and traces of ambient insect sounds come across in Georg Philipp Telemann's Symphony in G Major ("Grillen"; circa 1765), Benjamin Britten's *Two Insect Pieces* for oboe and piano (1935), and Tōru Takemitsu's "In an Autumn Garden" (1973). Miles Davis's trumpet evokes the buzzing of a hornet in "Frelon brun" ("Brown Hornet," 1961), and a synthesized cricket track trades off with a real cricket in New Boyz's "Cricketz" (2009). David Rothenberg jams with insects by playing an instrument while surrounded by a chorus of cicadas in nature, and David de la Haye composed pieces coupling human musicians playing alongside recordings of aquatic insects (for example, drumming insects in "Six-Limbed Drummer," 2022).

Jeff Claus and Judy Hyman, members of The Horse Flies, experimented with a friend when playing three fiddles using the same fingering, but in different tunings. They have also played different tunes with similar structures, but in different keys. They described the results to me this way:

> These approaches produced harmonic and rhythmic patterns we loved that morphed from harmonious to dissonant and back to harmonious within the rise, fall, and flow of the playing. It struck us as similar to the sound of insects when they get going, ebbing and flowing in undulating and shape-shifting waves and patterns on a warm humid night in a rural setting. We took to calling it "bug music" and even, at a certain point around the time of the release of the album Human Fly [1987], we came up with what we saw as a jokey description of our music, "neo-primitive, bug music."

Brief Imitations

Brief instrumental or vocal imitations of insect sounds are out there, if you listen carefully. "Flicka flicka flicka!" flutter caterpillar girl's wings, as Robert Smith of The Cure begins "The Caterpillar" (1984). "Zz—zz—zz—zz" buzzes Brazilian singer Simone imitating a cicada at the end of the song

"Cigarra" (from the 1978 album *Cigarra* [*Cicada*]). The cricket singing for the first twenty-five seconds of The Beatles' "Sun King" (1969) doesn't count, of course, because neither John, Paul, George, nor Ringo attempt to imitate the chirping insect, but the closing seven seconds of string staccato in Colt Ford's "Crickets" (2014) does, at least to my ear. Even more popular than cricket cameos are buzzing imitations, with an early example coming as waves of symphonic buzzing that open Ralph Vaughan Williams's "The Wasps" (1909), or the electronic shaver buzzing in Chubby Checker's "The Fly" (1961). Humans briefly buzz as flies in "Human Fly" (The Cramps, 1979), "Insects" (Oingo Boingo, 1982), "Wings Off Flies" (Nick Cave, 1984), "Feather Pluckn" (The Presidents of the United States of America, 1995), and "Mosquito" (Yeah Yeah Yeahs, 2013). No singer, however, has the stamina or vocal skill of a fly like Yoko Ono (coming next).

Full-Throttle Transformations

Mimicry has to be convincing in nature for mimics to survive. If natural selection were acting on humans as acoustic mimics of insects, the following musicians would prevail as humanity's survivors. I offer three categories of mimetic prowess: buzzing of insect wings, courtship calls, and the miscellaneous prize for mimicry of ant noises.

Yoko Ono has had a varied and influential career, but to this entomologist, the pinnacle was her recording of "Fly" (1971), in which for 22 minutes and 53 seconds she applied vocal acrobatics to screech out the buzzing of a house fly. Her husband, John Lennon, accompanied on guitar, added as a track played backward that also sounds eerily dipteran. "I was feeling like I was a fly," Ono later explained of her fly performance. One year later, The Who released (as a B side, ironically) "Wasp Man." Needing no more than a bare minimum of lyrics letting one know that a sting is imminent, the melody is driven by buzzing vocals. Add N to (X) harnesses electronica to produce the buzz of a wasp in "King Wasp" (1997). Here, too, lyrics are not the focal selling point of the music.

Before we jump from buzzing of insect wings to courtship calls, I will award my miscellaneous prize for unusual imitations to Lisa Schonberg's

"HVAC" (2022) for ant squeaking. Stridulation, like the rubbing of katydid's wings, involves one body part rasping against another to create noise, which can be used in a variety of functional contexts. Schonberg combines Jane Paik's voice imitating ant stridulations with field recordings from a leaf-cutter ant nest (*Acromyrmex* sp.), and percussion to evoke empathy for the ants, and to more closely relate to them. "Through mimicry we strive to express an approximation in mood, tone, and rhythm of the ants' sound, through our listening and in our physical response."

It is time for courtship, and our final category highlights the music of males serenading females. Cicadas are among the loudest of all insects, and when the trees are screaming with their calls, the mating melody can be overpowering, or inspiring, to human ears. The zenith of cicada imitations is vocally delivered by the Dong (Kam) people in Guizhou and Guangxi provinces, southwest China. Women erupt in a rapid-fire "nn-nn-nn-nn-nn-nn-nn-nn-nn" sung without instruments, but with layers of drones and complementary tones to recreate a chorus of males calling for females. An instrumental counterpart to the Dong peoples' chorus is a piece played by Lendungen Simfal and Ihan Sibanay, both T'boli people from a village in southwestern Mindanao, Philippines. "Utom kuleng helef" ("Call of the Cicada"; 1995, released 1997) uses bamboo zithers to capture the repetitive, rhythmic cicada choruses.

The most celebrated of the insect musicians are crickets. Brought indoors in ornate custom cages, jars, and gourds, crickets have deeply delighted human listeners since early Chinese antiquity, and appear in pieces of art, literature, and music. Despite historic adoration for crickets, vocal mimicry of cricket song may be rare, and the earliest clear example I have found is from Renaissance Europe. Josquin des Prez, purportedly in honor of or to make fun of a singer colleague with the last name Grillo, composed "El Grillo" ("The Cricket," 1505) for four voices. The voices playfully, rapidly rise and fall to simulate cricket stridulation. Using a larynx to mimic the stridulatory raspings of a cricket poses challenges, and sound and media artist Stephanie Loveless reported "in slowing them down, parsing their frequencies, and matching my voice to theirs as closely as I can—I hope to open myself to their respective worlds." In a work reminiscent of Yoko Ono's *Fly*, Loveless's *Cricket, Tree, Crow* (2013) is a vocal departure from anything human, altered into an eerie state of cricketdom.

Cricket Song in Space

Crickets and their relatives have been singing since at least the Triassic period, hundreds of millions of years before Josquin des Prez or Loveless orchestrated chirping impressions, and cricket song will likely last long after us. We have essentially ensured this by sending singing crickets into interstellar space. What follows is a story of hurling a satellite carrying a cricket song forever beyond our reaches, challenging us to consider our accomplishments, our ambitions, and how it will all end.

There are many ideas about what will become of us. Predictions suggest that in one billion years the sun will boil our oceans, and die out a few billion years later. What traces of us are left behind may be meager. We travel in a bubble of space containing our sun's plasma, and few things escape this heliosphere. The only tangible objects produced by humans to have broken free are *Voyager 1* and *Voyager 2*, space probes launched in 1977. *Voyager 1* managed this feat on 25 August 2012, and *Voyager 2* followed six years later. Now over 14.8 billion miles (23.8 billion km) away, *Voyager 1* is traveling on a trajectory to exit the Oort cloud toward the constellation Ophiuchus and, at 38,000 miles per hour (17 km/s), could reach the nearest star in 73,775 years, and not empty-handed. Thanks in great part to the efforts of astronomer Carl Sagan, *Voyagers 1* and *2* each carry a golden record containing greetings in 55 languages, 27 pieces of music, 115 images, and 35 sounds. The record carries a single image of an insect (*Ophion luteus*, ichneumon wasp, flying beside a flower), and one insect sound recording. The sound, though full of other ambient noises, like croaking frogs, is a chorus of crickets. On top of this nocturnal cacophony sings a lone serenading male (*Teleogryllus oceanicus*). The choices of what images and sounds to include were made with careful calculation, out of convenience, or because a tight deadline made certain options impossible. The ichneumon wasp joined a selection of images representing inhabitants of Earth, but nothing more specific about the insect or the image was intentional. The cricket recording was "intended to betoken the debut of vociferous life on earth," according to the project's creative director, Ann Druyan, and was recorded by neurobiologist Ron Hoy, conveniently working at the same university as Sagan. NASA ruled out an image of a pair of nude humans, and the sound of a kiss became the most difficult to capture.

Aside from the two odd "bottles" we launched "into the cosmic ocean," as Sagan referred to the *Voyager* spacecraft, so far we leave interstellar evidence of our existence as electromagnetic radiation, and nothing more. The first commercial radio broadcast, made on 2 November 1920 by Frank Conrad from Pittsburgh, Pennsylvania, has since traveled more than one hundred light years from Earth, so if an entity sits in the path of these waves and is prepared to receive them, they could be learning of Warren Harding's presidential victory at this very moment. A fleeting broadcast of this trivial human happening will likely travel through the universe without detection. Janna Levin, a cosmologist from the documentary *A Trip to Infinity* offers this personal take on cosmic extinction: "There will be a last sentient being, there will be a last thought."

This makes me think about the song of the long-extinct katydid, *Archaboilus musicus*. Our appreciation of music may have originated with insect sounds, from members of his order. That we are able to play his song gives us the tiniest window into the lost acoustics of our planet's past. We will leave our own fossils of sorts—aboard long-dead space probes, and electromagnetic evidence in the form of encoded radio waves. Will it be possible for others to intercept and decode such relics and listen to "Wasp Man" or "El Grillo" as we now listen to a recreation of the katydid's courtship call? If so, traces of how insects affected our culture may serve as our humble legacy, our final thought.

Melt by Erika Harrsch (2012)

Conclusion

The earth has spawned such a diversity of remarkable creatures
that I sometimes wonder why we do not all live
in a state of perpetual awe and astonishment.
—HOWARD ENSIGN EVANS, *LIFE ON A LITTLE KNOWN PLANET*

I am writing this in English, which owes its origins to Latin, Greek, Aramaic, Phoenician, all the way back to letters invented by a Semitic-speaking people. We have fragmented thoughts recorded by these people, but the very oldest, complete, decipherable sentence ever found in an early alphabetic

301

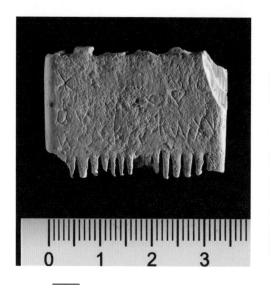

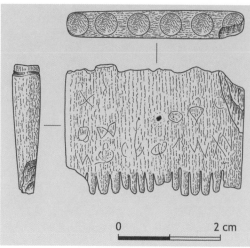

Reflectance transformation imaging photograph of ivory louse comb with incised sentence about lice

ABOVE RIGHT: Drawing of the louse comb. Letters one through nine go right to left (*bottom row*), and continue ten through seventeen, going left to right (when flipped; *top row*). "Louse" is represented by letters eight through ten.

script pertains to insects. Yosef Garfinkel, one of the archaeologists who helped decipher the seventeen letters forming seven words that transport us to the age of the Canaanites, put it this way: "This is the first sentence ever found in the alphabet." From the ancient ruins of Tel Lachish, Israel, comes an inscription made in 1700 BCE on a comb made from elephant ivory. The message is a plea: "May this tusk root out the lice of the hair and the beard. (ytš ḥṭ ḏ lqml śʿ[r w]zqt)"

"What were you expecting?" Garfinkel asked a journalist. "A love song? A recipe to make pizza?" It was inscribed on a louse comb, after all.

Our relationship with insects is ancient, deep, and conflicted. The ivory comb rooted out unwelcome parasites, and some of the great vectors of disease and destroyers of our crops are insects. Because we share the same habitats and compete for some of the same resources, we tend to bias our attention toward the fraction of insects that can pose problems for us. But of the 1.1 million described species of insects, less than half of one percent (five thousand) belong to this troubling fraction. And even these can offer surprising benefits to technology, medicine, and the arts as earlier examples show. The vast majority of insects play ecological roles we have only begun to understand, and insects provide basic and profound cultural services, though few of us recognize their involvement and influence.

A recurring revelation while writing this book emerges: Insects perform as the unseen actors behind much of what we do and who we are. Honey bee nests may have supplied our ancestors the energy needed to beat out competition and fuel our expanding brains, and today humans traditionally use well over 1,600 species of insects for food and over 80 species for medicine. Insect sounds may have instilled in us our love of rhythm and music. Silk, wax, honey, cochineal, lac, and other insect products have steered economic trade, which facilitated the spread of language and ideas, and transformed art, clothing, and technology. By one author's count, ten of the fifty animals that changed the course of history were insects (full disclosure: the author is a vertebrate who included *Homo sapiens* among the fifty). Another author cites twelve insects among fifty-four animals as key inspirations for engineers (and yes, one of the other animals is *Homo sapiens*). Steven Kutcher, insect wrangler for movies seen by millions, estimates that one in every three films includes insects. By almost every measure, insects play an outsize role in shaping our existence, so what happens when our actions lead to their decline?

Some are calling the geological epoch in which we live the Anthropocene—or "human era"—named after us, in recognition of our indelible impact on life and the planet. Our rapid-fire destruction of habitats is resulting in biodiversity losses, which is especially relevant to insects because of their ubiquity and unmatched diversity among animals. This makes it relevant to us, in part because species with untold roles in nature and unknown potential to positively influence our lives are going extinct daily. By virtue of our short lives and shorter memories, we tend to forget what we have lost, and lose sight of what makes for a healthy world.

In a future age, I envision a curious creature exploring layers of sediments, as a fossil hunter does today. Eventually, the creature uncovers a sliver of earth, a stratum deep under the surface with hints of our remains. Shallower and narrower than the preceding ages of dinosaurs or fishes, ours

Wallpaper with endangered species, including eighteen insect species (*Red list wallpaper KYOTO 2015*; detail of inkjet print, 2021). To Asuka Hishiki, "This piece is a reminder to myself that we can be blind to the devastating problem. Like 'wallpaper,' it is there in front of us, but we tend to ignore it as vague background decoration."

might stand out as peculiarly divisive and dramatic for its relative brevity. Below our layer, the creature might find teeth and fragments of giants, but then we arrive and the giants largely vanish. Evidence would show a wave of agriculture and an age of industrialization. Plastic appears for the first time, along with relics of manufactured mixed media. The creature might expand their study, and find that wherever we appear in the fossil record, rapid changes follow. Above our layer, closer to the time of our curious creature's fossil excavations, there are still remnants of other species, but the traces are scant and scattered. We enter stage right, and a curtain falls on our fellow actors as we exit stage left.

As bleak as the just-detailed scenario sounds, there is ample room for hope. Adversity can force us to reassess the status quo, and to challenge our destructive behavior.

Marlène Huissoud's *Please Stand By* (2021; unfired clay, natural binders, wood) are sculptures to be inhabited by urban insect pollinators in need of a home.

Frank Oppenheimer, the particle physicist whose older brother is often referred to as the "father of the atomic bomb," wanted to find ways to help humanity in response to the bomb. Branded a communist for his affiliation with the communist party during the Great Depression, Oppenheimer was denied a passport and could not find a job in physics. When faced with "real-world" obstacles, he would declare, "It's a world we made up, so we can remake it anyway we like." This philosophy, and his desire to improve human relations, are what allowed him to found the Exploratorium, a community museum and learning lab in San Francisco, and to combat entrenched thinking of his time. So, too, Martin Luther King Jr. recognized the "fierce urgency of now," and helped lead the civil-rights movement; Mahatma Gandhi applied nonviolent resistance against colonialism; and Rachel Carson challenged corporate lies to spark the environmentalism movement. The same philosophy of fighting against deep-seated wrongs could serve us well when faced with laws, corporations,

or governments that would have us believe we must embrace an unsustainable existence. No social movement is inevitable, and neither is our fossil-hunting fable. For the first time in human history there is a global reckoning of our actions as a species, and people are reconsidering how we obtain energy, how we farm, what we eat, how dark we keep our night sky, and what it means to have a beautiful yard. A single person can make a difference, and each of us can change the world in ways that promote life rather than hinder it.

What gives me optimism? When I share insect stories with children, their eyes light up, their hands shoot in the air, and their desire to recount personal tales of how insects appear in their lives is electric. When I want to find comfort, I huddle down in front of the smallest patch of earth and patiently watch for evidence of activity, of vibrant life. Even a tiny area can

house as many organisms as there are humans on the planet . . . or who have ever *been* on the planet. In this patch, I see insect societies with elaborate architecture, predators with specialized legs for injecting venom, escape artists with springing tails, chemical defenses, chemical communication, camouflage, warning coloration, cooperation, foldable wings, and apparatus for sensing the world beyond our experience or understanding, knowing that discoveries await and lessons abound. I see a tree connecting millions of lineages, including ours. This gives me hope, and joy.

ACKNOWLEDGMENTS

Every human who has ever lived for an appreciable length of time has formed a personal relationship with insects. Among the billions of us, there are those who recognize the beauty or utility of our neighbors, and choose to explore their relationship with ingenuity and creativity. My greatest joy while writing this book has been to find and meet so many inspired people who explore entomology in their individual, personal, and professional ways. I have only managed to reach a tiny subset of the world's living luminaries who embrace insects in their lives, and their thoughts, inventions, research, writings, and art have made this project a perpetual thrill for me. I am deeply grateful to all, and hope that my inclusion of their contributions gives readers incentive to explore more.

My family gives me strength, motivation, love, and unbridled support. You will find my mother's art, my father's definition of art, my twin brother's cautionary tale, and my sister's lutherie in these pages. My eight-year-old son, Rivyn, is an unparalleled editor, and created a variant cover for this book that features a deep-sea anglerfish levitating above a sting-bearing castle (to be used in a future edition). Dosha, my wife and my greatest reader, carefully and constructively critiqued my writing, challenging me in the most helpful ways. Had she been my ghostwriter, the prose would have entranced and consumed you.

Some painters transform the sun into a yellow spot,
others transform a yellow spot into the sun.
—PABLO PICASSO

To the artists! Elizabeth Jean Younce, whose work enthralls me, created original pieces for the cover, title page spread, and opening each section.

My mother, Karen Anne Klein, not only gave birth to me, but is my closest collaborator. She produced drawings for the contents section and beyond. Alison Rogers Napoleon, Arno Klein, and Bill Harris honored me by creating original written works to open each of the book's sections. So many artists shared their insectile visions with me, and now with you.

Charles Hogue formalized the study of cultural entomology, and left a foundation for a truly interdisciplinary field of study. Tom Seeley, honey bee mentor, collaborator, and friend, shared his experience as a writer, as did Michael Engel and Raghavendra Gadagkar. The Pupating Lab, Arno Klein, Steven Kutcher, and my entomology class (particularly Lily Gritzmaker and Cedar Sekorski) gave helpful feedback on early drafts. Carleton College "extern," Aaron Berkowitz, suggested the section subtitles. Erin Roup assisted with reference formatting. The University of Wisconsin–La Crosse granted me a sabbatical, which strangely coincided with a pandemic.

During the course of writing, I have corresponded with people inside the CIA, Museum of Jurassic Technology, Reserve Bank of Fiji, Bruce Lee

BARRETT, the cover of Syd Barrett's final solo album (1970), featuring a drawing produced years earlier by Barrett. Syd Barrett was a songwriter and musician who co-founded Pink Floyd (Syd Barrett Music Ltd.).

Foundation, and contributors from all continents but Antarctica (too few insects there). For information or kind assistance, I thank Marta Areia, Don Ashby, Lucinda Backwell, Ian Barrett, Robert Bégouën, Megan Biesele, Tierney Brosius, Jennifer Brown, Jim Carpenter, Caroline Chaboo, Jeff Claus, Joe Coelho, Tobias Deschner, Karen Duffek, Whitney Fisher, Yosef Garfinkel, Jun-Jie Gu, John Hansen, Bug Hartsock, Eva Hausam, Arnold van Huis, Judy Hyman, Julia Kennedy, Seth King, Gene Kritsky, Seth Kroll, Julie Lesnik, Paul Loasby, Bill Logan, Léona Béatrice Martin-Starewitch, Manuel Bea Martínez, Rosa Mei, Jason Nargis, Michael Nylan, Mikkel Pedersen, M. Alejandra Perotti, Simon Peers, Bob Peterson, Christopher Philipp, Simone Pika, Erin Powell, Deborah Rudolph, Lisa Schonberg, Claire Spottiswoode, William Stobb, Eugene Tssui, Lola Wegman, Sydnie Wilson, and all of the artists and scientists who have graciously allowed me to represent their work. Thank you to Timber Press for giving me a chance to write my first book, including Will McKay for reaching out to me, Mary Cassells, Sarah Milhollin, Matthew Burnett, Sarah Crumb, and Ryan Harrington for seeing this come to life.

PHOTO AND ILLUSTRATION CREDITS

Page 2: Elizabeth Jean Younce

Page 5: Amey Yun Zhang

Pages 6–7: Karen Anne Klein

Page 10: Uruk period (4000–3100 BCE) or later, excavated from Shara Temple in Tell Agrab, Iraq (seal: 0.8 [H] × 0.9 in. [D] / 22 x 25 mm; 1935–1936). Institute for the Study of Ancient Cultures, West Asia and North Africa; University of Chicago. Accession Number: A18117 and C8903. Photo by Barrett Klein.

Page 11: Discovered in 1928. Paris, Bégouën Collection, Musée de l'Homme. Accession Number: 55330004. Photo by Robert Bégouën, Association Louis Bégouën.

Page 15: *Biodiversity*, etching plus color pencil. Barrett Klein (2008).

Page 18: Insectology Food for Buzz, *Hoverfly* by Matilde Boelhouwer. Photo by Janneke van der Pol.

Page 21: Photo by Wendy Taylor CBE

Page 23: Elizabeth Jean Younce (2023)

Page 26: The Minneapolis Institute of Art/ *Forest Coat* by Jon Eric Riis (24 × 64 in. / 61 × 163 cm; 2005). The Ethel Morrison Van Derlip Fund. Accession Number: 2004.193, image used courtesy of the artist.

Page 29: (*top*) The Metropolitan Museum of Art, New York/*Emperor Justinian Receiving the First Imported Silkworm Eggs from Nestorian Monks*, Plate 2, from The Introduction of the Silkworm (Vermis Sericus), engravings by Karel van Mallery, after Jan van der Straet (7.8 × 10.6 in. / 19.9 × 26.8 cm;

circa 1595). The Elisha Whittelsey Collection, The Elisha Whittelsey Fund (1949). Accession Number: 49.95.869(3). (*bottom*) The Metropolitan Museum of Art/*The Reeling of Silk, Plate 6*; gift of Georgiana W. Sargent, in memory of John Osborne Sargent (7.2 × 10.6 in. / 18.3 × 26.8 cm; 1924). Accession Number: 24.63.492.

Page 30: The Art Institute of Chicago/Clarence Buckingham Collection. No. 2 from the twelve-part series *Women Engaged in the Sericulture Industry* (*Joshoku kaiko tewaza-gusa*), by Kitagawa Utamaro I; woodblock print, ink, and color on paper (15.3 × 10.3 in. / 38.9 × 26.2 cm; circa 1798–1800). Accession Number: 1925.3247.

Page 31: The Art Institute of Chicago/Clarence Buckingham Collection. Accession Number: 1925.3251.

Page 33: From the renwu (人物) section of the Chinese encyclopedia Sancai tuhui (三才圖會, 1609 edition), chapter 10, pages 14a–b (text block 16¼ × 11 in. / 41.3 × 28 cm); courtesy of the East Asian Library of the University of California, Berkeley.

Page 35: Wellcome Collection, Public Domain Mark/*Micrographia: or some physiological descriptions of minute bodies made by magnifying glasses. With observations and inquiries thereupon. By Robert Hooke* (1665).

Page 36: Tera Galanti

Page 39: Cleveland Museum of Art, John L. Severance Collection/*Sericulture (The Process of Making Silk)* (third section: 10.9 in. [H] / 27.6 cm). Accession Number: 1977.5. Attributed to Liang Kai 梁楷 (early 1200s).

Page 40: Xu Bing, *Silkworm Book: The Analects of Confucius* (2019). Book, silk (1¾ [H] x 20½ [W] x 16½ in. [D] / 4.4 x 52 x 42 cm). © Xu Bing Studio.

Page 41: Kazuo Kadonaga, Kadonaga Studio (1986)

Page 42: Jen Bervin Studios

Page 43: Biodiversity Heritage Library/*Brehms Tierleben. Allgemeine kunde des Tierreichs*, by Alfred Edmund Brehm. Volume:Bd.2 (1915). Image by P. Flanderky, H. Morin, G. Müßel, or E. Schmidt. Made available through Biodiversity Heritage Library and the Smithsonian Libraries and Archives.

Page 44: (*left*) *Surreal Banquet: Lymantria dispar dispar*, collage within handmade book by Karen Anne Klein (5½ × 4½ in. / 14 × 11.4 cm; 2022). (*right*) Rare Book Division, The New York Public Library Digital Collections. The Trouvelot Astronomical Drawings Atlas, *The November Meteors: As Observed Between Midnight and 5 o'clock a.m. on the Night of November 13–14, 1868* (1881–1882).

Page 46: Internet Archive/*Contributions to the systematics of the caddisfly family Limnephilidae (Trichoptera) I*, illustrated by Anker Odum (1973) in book by Glenn B. Wiggins (page 18). Royal Ontario Museum.

Page 48: © AKI INOMATA/*girl, girl, girl . . .* (2012/2019), courtesy of Maho Kubota Gallery

Page 50: Minneapolis Institute of Art/*Man's Under Kimono (Nagajuban) with Spider and Spiderweb* (1920s–1930s). Gift of John C. Weber. Photo by Barrett Klein.

Page 51: (*left*) Courtesy of The Field Museum. Tall mask (24.2 in. / 61½ cm). Accession Number: 133088. (*right*) Mask minus

webbing (4.9 × 5.1 in. / 12½ × 13 cm). Accession Number: 133057. Acquired 1909–1913; J. N. Field Expedition, A. B. Lewis, collector. Field Museum of Natural History. Photos by Barrett Klein.

Page 52: Northwestern University's Charles Deering McCormick Library of Special Collections. Portrait of Philippine Welser (oval: 3 × 4 in. / 7.6 × 10.2 cm); courtesy of Charles Deering McCormick Library of Special Collections and University Archives, Northwestern University Libraries. Photo of piece with backlighting, courtesy of Jason Nargis.

Page 54: Courtesy of Simon Peers from *Opuscoli scientifici d'entomologia di fisica e d'agricoltura dell'abate*, by D. Raimondo Maria de-Termeyer; collection of Simon Peers

Page 55: © Victoria and Albert Museum, London, acquired 2011 (55½ × 51.2 in. / 141 × 130 cm; 1889)

Page 60: The Metropolitan Museum of Art, New York/Gift of Mrs. Otto H. Bacher (11.4 × 8.3 in. / 29 × 21.1 cm; 1938). Accession Number: 38.14.2.

Pages 62–63: (*left*) Hive helmet (15.7 × 23.2 × 11.8 in. / 40 × 59 × 30 cm; 2022); (*right*) Hive helmet detial (16½ × 8.3 × 11.4 in. / 42 × 21 × 29 cm; 2020). Photos by Barrett Klein (2010).

Pages 64–65, 66 (*top*): Photos by Gene Kritsky, fused and modified to form a single composite image by Barrett Klein (24 × 6½ in. / 61 × 16.4 cm)

Page 66 (*bottom*): Photo by Livioandronico2013 (2014), Wikimedia/Creative Commons

Page 67: Photographs and tracing by Manuel Bea Martínez (University of Zaragoza, Spain)

Page 68: (*left*) Gene Kritsky (length of hunter: 1½ in. / 3.7 cm; branches to end of ropes, which continue beyond what is shown in the images: 21.6 in. / 55 cm), photo by Gene Kritsky; (*right*) tracing by Barrett Klein

Page 70: Photos by Claire Spottiswoode

Page 71: Photo by David Lloyd-Jones

Page 72: (*left*) Dr. Dale Pollett; (*right*) photo by

Erin Powell (Alachua County, Florida; 2023)

Page 73: Photo by Dr. M. Alejandra Perotti, Museo Chileno de Arte Precolombino, Santiago, Chile

Page 77: State Museums in Berlin, Kupferstichkabinett/Pieter Bruegel the Elder (ca. 1525–1569), pen and brown ink (12.2 × 8 in. / 30.9 × 20.4 cm). Accession Number: KdZ713. Public Domain Mark.

Page 81: Courtesy of The Field Museum of Natural History (lac staff: circa 11 in. / 28 cm long; 793 g). Cat. #48485; stick: #48484. Photo by Barrett Klein.

Page 86: Photo by Erin C. Powell (Alachua County, Florida; 2023)

Page 87: Photo by Barrett Klein (Pedernales Falls, Texas; 2004)

Page 89: Created by Jennifer Angus for "Insect Dreams Cabinet" (collection of books by thirty-seven artists curated by Karen Anne Klein and Barrett Klein in 2019)

Page 90: Aslı Çavuşoğlu, *Red/Red* (2015). Installation view of Saltwater, 14th Istanbul Biennial. Photo by Sahir Uğur Eren.

Page 91: Art Institute Chicago/*The Bedroom* by Vincent van Gogh (1889); oil on canvas (29 × 36.3 in. / 73.6 × 92.3 cm). Helen Birch Bartlett Memorial Collection. Accession Number: 1926.417. Creative Commons Zero.

Page 93: British Library/Magna Carta, burnt copy with the seal attached. (To see text rendered visible via multispectral imaging, visit the British Library's site.) Cotton Charter XIII 31 A. Public Domain (except in the UK).

Page 95: Barrett Klein

Page 99: Biodiversity Heritage Library/*Fabre's Book of Insects*, retold from Alexander Teixeira de Mattos's translation of Fabre's *Souvenirs entomologiques*, Plate 8, by Edward Julius Detmold. Image in the Public Domain.

Page 100: Google Books/*Versuche und Muster,*

ohne alle Lumpen oder doch mit einem geringen Zusatze derselben, Papier zu Machen (1765), Volume 1, by Jacob Christian Schäffer

Page 101: Alastair and Fleur Mackie (house: 92½ × 49.2 × 39.4 in. / 235 × 125 × 100 cm; 2008); courtesy of Alastair and Fleur Mackie

Page 107: Karen Anne Klein (10½ × 10½ in. / 26.7 × 26.7 cm; 2009), photo by Eric Law/ShootMyArt.com

Page 110: Photos by Caroline S. Chaboo (‡Nlundi Village, Aha hills; 2007)

Page 111: (*left*) Photo by Caroline S. Chaboo, with appropriate permissions obtained; (*right*) courtesy of the Division of Anthropology, American Museum of Natural History, Western Ngamiland/Dobe, Botswana (quiver: 21 in. / 53.3 cm; arrows: 13½–21 in. / 34.3–53.3 cm). Accession Number: 90.2/6001 A-VV. Photo by Barrett Klein.

Page 113: Tobias Deschner/Ozouga Chimpanzee Project (still image extracted from video taken by Tobias Deschner in 2021 as part of the Ozouga Chimpanzee Project)

Page 115: Courtesy of the Division of Anthropology, American Museum of Natural History. (*top left*) with macaw, parrot, and rooster feathers, Accession Number: 40.1/3235; (*bottom left*) Accession Number: 40.1/3241; (*wasps on right*) Accession Number: 40.0/6352. Photos by Barrett Klein.

Page 121: Wikimedia/Siga CC BY-SA 3.0

Page 123: Cornelia Hesse-Honegger, *Garden Bug from Küssaberg* (1991); watercolor (16½ × 11.7 in. / 42 × 29.7 cm). © 2024 Artists Fights Society (ARS), New York; ProLitteris, Zurich.

Page 125: The Metropolitan Museum of Art, New York/The Howard Mansfield Collection; gift of Howard Mansfield (1.45 and 1.3 in. / 3.7 and 3.3 cm; 1936). Accession Number: 36.120.462a–b.

Page 129: Courtesy of the Reserve Bank of Fiji (image modified with the word "SPECI-MEN," as required by the Reserve Bank of Fiji)

Page 130: Photo by Gene Kritsky

Page 132: Photo by Joseph Yoon; credit to Brooklyn Bugs

Page 133: Photos by Lucinda Backwell and Francesco d'Errico (line segments indicate breaks; scale bar [horizontal line segment]: 0.39 in. / 1 cm)

Page 135: Isabella Kirkland, oil and alkyd on polyester over panel (36 × 48 in. / 91.4 × 121.9 cm). Key to the sixty-six species can be found at IsabellaKirkland.com/edibles.

Page 141: Barrett Klein (1993; colored in 2024)

Page 149: Wikimedia/Diego Delso, https://delso.photo/, CC BY-SA 4.0

Page 151: Eleanor Lutz

Page 153: The Metropolitan Museum of Art, New York/Gift of Robert Hatfield Ellsworth, in memory of La Ferne Hatfield Ellsworth (7 in. [T] / 17.8 cm; image cropped; 1986). Accession Number: 1986.267.78.

Page 155: Photo by Piotr Naskrecki

Page 157: Nao Kimura (1994), © Yanagi Studio

Page 158 (image A): La Beauté en Avignon/Angel, installation by Nick Knight (2000). Image courtesy of Nick Knight and Lee Alexander McQueen.

Page 158 (image B): Suze Woolf

Page 158 (image F): The J. Paul Getty Museum, Los Angeles/Stag Beetle. Accession Number: 83.GC.214. Digital image courtesy of Getty's Open Content Program.

Page 158 (image G): Wikimedia/Egyptian Museum of Cairo, Egypt. Photo by Wael Mostafa, Wikimedia Commons.

Page 158 (image H): The Metropolitan Museum of Art, New York/The Michael C. Rockefeller Memorial Collection; purchase of Nelson A. Rockefeller; gift (1968). Accession Number: 1978.412.206.

Page 158 (image L): Studio Marlène Huissoud

Page 158 (image M): The Metropolitan Museum of Art, New York/Gift of Abby Aldrich Rockefeller (1937). Accession Number: 37.92.13.

Page 158 (image N): The Metropolitan Museum of Art, New York/Purchase of Lila Acheson Wallace, Drs. Daniel and Marian Malcolm, Laura G. and James J. Ross, Jeffrey B. Soref, The Robert T. Wall Family, Dr. and Mrs. Sidney G. Clyman, and Steven Kossak; gifts (2008). Accession Number: 2008.30.

Page 159 (image C): Steven R. Kutcher

Page 159 (image D): The Metropolitan Museum of Art, New York/Purchase of Alice and Evelyn Blight and Mrs. William Payne Thompson; gift (2003). Accession Number: 2003.303.

Page 159 (image E): Special Collections Research Center at NC State University Libraries/Insectes (1920), by Emile-Allain Séguy, 1877–1951. Accession Number: QL466.S49.

Page 159 (image I): The Metropolitan Museum of Art, New York/H. O. Havemeyer Collection, Bequest of Mrs. H. O. Havemeyer (1929). Accession Number: 29.100.370.

Page 159 (image J): The Metropolitan Museum of Art, New York/The Jules Bache Collection (1949). Accession Number: 49.7.19.

Page 159 (image K): The Louvre Museum/Victory of Samothrace. Accession Number: Ma 2369. Photo by Lyoko.88, Wikimedia Commons.

Page 159 (image O): The Metropolitan Museum of Art, New York/Rogers Fund (1959). Accession Number: 59.92.

Page 159 (image P): Musée d'Orsay, Dist. RMN-Grand Palais/Arrangement en gris et noir n°1, by James Abbott McNeill Whistler (1871). Accession Number: RF 699. Wikimedia Commons, Public Domain. Photo by Patrice Schmidt.

Page 162: Courtesy of the Division of

Anthropology, American Museum of Natural History (5½ × 3½ × 1.8 in. / 14 × 9 × 4½ cm; Iquitos, Peru). Accession Number: 40.0/3509. Photo by Barrett Klein.

Page 163: Tomoaki Inaba

Page 165: Museum of Jurassic Technology/Eva Hausam, curator. Photos by Lola Wegman.

Pages 166–167: Field Museum of Natural History. Page 166 (*top left*): 29.3 in. / 74½ cm; each head and horn: 1 in. / 2½ cm (acquired 1968–1969); P. David Price, collector; Accession Number: 190826. Page 166 (*bottom left*): seven loops, each: 21.7 in. / 55 cm; each leg segment: 0.16–0.20 in. / 4–5 mm (acquired 1968–1969); P. David Price, collector; Accession Number: 190825. Page 166 (*top right*): 15½ in. / 39.4 cm long; each leg segment: 0.12–0.20 in. / 3–5 mm (acquired 1909–1913); J. N. Field Expedition, A. B. Lewis, collector; Accession Number: 137138. Page 167 (*bottom*): circumference: 11½ in. / 29.2 cm; each beetle: 1 in. / 2½ cm long; collected by Suzanne Kaufman (1960s–1980s); Field Museum of Natural History; Accession Number: 351986. Page 167 (*top*): each body: 0.16–0.20 in. / 4–5 mm; Accession Number: 28750-1. Photos by Barrett Klein.

Page 168: *Eumargarodes laingi*, family Margarodidae (0.06–0.12 in. / 1½–3 mm long). Photo by Erin C. Powell (Okaloosa County, Florida; 2022).

Page 169: Akihiro Higuchi

Page 170: Ruth Marsh

Page 171: Courtesy of the artist, Maria Fernanda Cardoso

Page 172: Donna Conlon, courtesy of the artist and DiabloRosso (Panama), and Espacio Mínimo (Madrid)

Page 173: Catherine Chalmers

Page 175: (*top left*) Tessa Farmer, courtesy of New Art Gallery Walsall, and Tessa Farmer, Natsko Seki, and Barrett Klein, respectively; (*bottom left*) photo by Natsko Seki

Page 179: Elizabeth Jean Younce

Page 183: Alexey Polilov (scale bar: 200 μm = 0.0079 in. / 0.02 cm)

Page 187: (*top left*) Minneapolis Institute of Art/ The James Ford Bell Foundation Endowment for Art Acquisition and gift of funds from Siri and Bob Marshall (28 × 24 × 13¾ in. / 71.1 × 61 × 34.9 cm). Accession Number 2012.31.1a–c. (*top right*) The Metropolitan Museum of Art, New York/ Gift of Etsuko O. Morris and John H. Morris Jr. (13 × 25¾ × 15½ in. / 33 × 65.4 × 39.4 cm; 2018). Accession Number: 2018.833.9a–d. (*bottom*) Samurai Museum, Berlin. Photo by Manfred Sackmann.

Page 189: Alamy/Codex Atlanticus, painting, verso (1051); https://codex-atlanticus. ambrosiana.it. Photo © Veneranda Biblioteca Ambrosiana/Mondadori Portfolio

Page 191: Young Min Song and colleagues (lens diameter: 0.006 in. / 0.16 mm; hemisphere: 0.63 in. / 1.6 cm; 2013)

Page 197: (*top*) Flickr/Jean-Pierre Dalbera CC BY 2.0. (*middle*) Jean-Pierre Dalbéra (2008), Wikimedia Commons. (*bottom*) Bokwang Song et al. *Reproducing the hierarchy of disorder for Morpho-inspired, broad-angle color reflection. Scientific Reports* 7 (2017), 46023; doi: 10.1038/ srep46023.

Page 199: Photo by Shuki Kato

Page 203: Photo by Barrett Klein (Ithaca, New York; 2007).

Page 210: (*left*) Photo by David Hu and John Bush, MIT; (*right*) photo by Kyujin Cho and Hoyoung Kim, Seoul National University (2015)

Page 211: Recorded by David Hu and colleagues, photo by David Hu and John Bush, MIT (2003)

Page 212: Yang et al. (2021), photos by Mable Fok (2011)

Page 215: CIA/Insectothopter (2.4 × 3½ × 0.6 in. / 6 × 9 × 1½ cm; 1970s). CIA photos are in the Public Domain.

Page 218: Wyss Institute at Harvard University

Page 223: Bill Logan, stonefly nymph (2 in. [L] / 5 cm; 1997); western golden stonefly (pages 100–101; 1997–1998); mayfly nymph (pages 88–89; 1998)

Page 225: (*top left*) The Metropolitan Museum of Art, New York/Purchase of Edward S. Harkness; gift; scarabs, blue glazed steatite (0.39 × 0.87 × 0.59 in. / 1 × 2.2 × 1½ cm; 1926). Accession Number: 26.7.713. (*top right*) The Metropolitan Museum of Art, New York/Rogers Fund; ring bezel (0.98 × 0.59 × 0.24 in. / 2½ × 1½ × 0.6 cm; 1936). Accession Number: 36.3.16. (*bottom*) Calouste Gulbenkian Foundation, Lisbon, Calouste Gulbenkian Museum (10.4 × 9.1 in. / 26½ × 23 cm). Photo by Catarina Gomes Ferreira.

Page 227: (*top*) Ricky Boscarino (aka Ricky of Luna Parc), Hieronymus Beetle (silver and bronze; 5¼ in. / 13 cm long; 1998). (*bottom*) Elizabeth Goluch, Violin Beetle, *Ode to Joy* (10½ in. / 26.7 cm long; 2008). Photo by Steven Kennard.

Page 228: Haruo Mitsuta (JIZAI *Chalcosoma chiron* and JIZAI *Mesotopus tarandus*)

Page 229: Noriyuki Saitoh

Page 231: Photo by Barrett Klein at the American Museum of Natural History

Page 232: (*left*) American Museum of Natural History, photo by Denis Finnin; (*right*) Mantis photo by Barrett Klein (urethane, hotmelt glue, wire, paint, and acuformed plastic mantis wings; 1998)

Page 233: (*top*) Oliver Zauzig, © Humboldt University of Berlin, Institute of Biology, Comparative Zoology group (4.1 in. [L] / 10.4 cm). Accession Number: 42054. (*bottom*) Whipple Museum of the History of Science, University of Cambridge (17.3 in. [L] / 44 cm; 1878). Accession Number: Wh.5181.

Page 234: Museum für Naturkunde, Berlin (100x actual size, 31½ × 19.7 in. / 80 × 50 cm; 1940s). Photo by Barrett Klein.

Page 235: The Ware Collection of Blaschka Glass Models of Plants, Harvard

University Herbaria/Harvard Museum of Natural History © President and Fellows of Harvard College. Glass bees (0.51 in. / 13 mm); single enlarged bee (1.65 in. / 42 mm). Hillel Burger Photographs (1981).

Page 239: Walter Tschinkel (19.7–23.6 in. [T] / 50–60 cm; 2012), photo by Charles F. Badland

Page 241: Images courtesy of Eugene Tssui

Page 244: Shutterstock/chettarin (*top*); Shutterstock/OG LOC VISUALS (*bottom, left*); Wikimedia/Derzs Elekes Andor CC BY-SA 3.0 (*bottom, right*)

Page 247: Elizabeth Jean Younce

Page 250: Lea Bradovich, pastel hand-colored etching (16½ × 12 in. / 41.9 × 30½ cm)

Page 253: Shon Kim. BOOKANIMA: Praying Mantis (2022) and BOOKANIMA: Tang Lang Quan; Shaolin Quan; Chinese Kick (2018).

Page 255: Photo shoot on Palos Verdes Beach, California (1966). The Bruce Lee name, image, likeness, and all related indicia are intellectual property of Bruce Lee Enterprises, LLC. All rights reserved. www.brucelee.com.

Page 259: Division of Anthropology, American Museum of Natural History (Hopi Reservation, Arizona). Accession Number: 50/9321. Photo by Barrett Klein.

Page 261: Courtesy of UBC Museum of Anthropology, Vancouver, Canada. Made by Joe Seaweed (bumblebee: 19.7 × 7.1 × 5½ in. / 50 × 18 × 14 cm; mosquito: 11 × 7.9 × 13 in. / 28 × 20 × 33 cm). Accession Numbers Nb3.1362 and A6213. Photo by Jessica Bushey.

Page 262: The Metropolitan Museum of Art, New York/H. O. Havemeyer Collection, Bequest of Mrs. H. O. Havemeyer (8.3 × 7.2 in. / 21.1 × 18.4 cm; 1929). Accession Number: JP2152.

Page 263: Courtesy of The Field Museum (8 feet [D] / 2.4 m). Accession Number: 138910.

Page 264: The University of Sydney/Jennifer

Napaljarri Lewis, Warlukurlangu Artists of Yuendumu. Photo by Louise M. Cooper.

Page 265: (*top*) Paul B. Goode (2013); (*bottom*) Martin Liebermann, www.martin-liebermann.de.

Page 267: Museum of Fine Arts Boston/Ernest Wadsworth Longfellow Fund (11.6 × 7 in. / 29½ × 17.8 cm). Accession Number: 2013.1667. Photo by Benjamin Joseph Falk (1901).

Page 269: Milena Sidorova, international choreographer, Young Creative Associate with Dutch National Ballet (2000)

Page 272: Barbara Walton (24 × 24 in. / 61 × 61 cm; 2011)

Page 274: Photo by Barrett Klein (2005)

Page 277: Karen Anne Klein, watercolor and color pencil (15¼ × 13¼ in. / 38.7 × 33.7 cm); photo by Eric Law/ ShootMyArt.com

Page 280: *Ant Queen Aria*, still image from The Colony performance, Anna Lindemann (2019)

Page 282: *Metamorphosis* by Ovid Works, art by Matylda Kozera

Page 285: (*top*) Courtesy of Ink Dwell. Photo of mural by Ben Zach and shoe by Le Mondeur and Ink Dwell; both courtesy of Ink Dwell.

Page 286: Photo by Polo Aguila (aka Poloide93), Creative Commons

Page 287: (*top*) firstVIEW, Ready-to-Wear Runway Collection, Fall/Winter; (*bottom*) Karen Anne Klein

Page 288: Bug Hartsock

Page 289: Barrett Klein, with silk tunic and wings sewn by Sienna Miller (2017 and 2018); photo of silk outfit by Dosha Klein

Page 290: (*left*) Dave Beatty; (*right*) Jennifer Angus

Page 291: (*top*) Laurent Seroussi (86.6 × 52 in. / 220 × 132 cm; 47.2 × 55.1 in. / 120 × 140 cm); (*bottom*) Laurent Gauthier (15.4 × 9.8 × 2.2 in. / 39 × 25 × 5½ cm). Photo by Philippe Paget.

Page 293: Charles E. Burchfield, 1893–1967, *Gateway to September* (1945–1956). Watercolor on joined paper (42 × 56 in. / 106.7 × 142.2 cm). Hunter Museum of American Art, Chattanooga, Tennessee; gift of the Benwood Foundation. Reproduced with permission of the Charles E. Burchfield Foundation.

Page 294: Photos and illustration courtesy of Jun-Jie Gu, Fernando Montealegre-Z, and Dong Ren

Page 301: Erika Harrsch; acrylic, ink, collage (71 × 47 in. / 180 × 120 cm; 2012)

Page 302: (*left*) Dafna Gazit, senior photographer, Israel Antiquities Authority; (*right*) courtesy of Institute of Archaeology, the Hebrew University of Jerusalem

Page 304: Asuka Hishiki

Page 305: Bernardo Figueroa

Page 306: Slinkachu

Page 307: Slinkachu, commissioned by the Trustees of the Natural History Museum, London

Page 309: Syd Barrett Music LTD. Reproduced by kind permission.

Page 354: Image courtesy of Dear Climate

Page 368: Karen Anne Klein, photo by Eric Law/ShootMyArt.com

RELATED RESOURCES

A TINY ENTOMO-SAMPLER

Entomophilia (Love of Insects)

Abbott JC, Abbott KK. 2023. *Insects of North America*. Princeton University Press. Princeton, NJ.

Eisner T. 2003. *For Love of Insects*. Harvard University Press. Cambridge, MA.

Elliott L, Hershberger W. 2006. *The Songs of Insects*. Houghton Mifflin Co. New York, NY.

Engel M. 2018. *Innumerable Insects: The Story of the Most Diverse and Myriad Animals on Earth*. Union Square & Co. New York, NY.

Grimaldi D, Engel M. 2005. *Evolution of the Insects*. Cambridge University Press. Cambridge, UK.

Hoyt E, Schultz T. 1999. *Insect Lives*. Mainstream Publishing. Edinburgh, Scotland.

Hubbell S. 1993. *Broadsides from the Other Orders: A Book of Bugs*. Random House. New York, NY.

Raffles H. 2010. *Insectopedia*. Vintage. New York, NY.

Stawell R. 1935. *Fabre's Book of Insects: Retold from Alexander Teixeira de Mattos' Translation of Fabre's "Souvenirs Entomologiques."* Tudor Publishing Co. New York, NY.

Insect Declines

Goulson D. 2021. *Silent Earth: Averting the Insect Apocalypse*. Harper Perennial. New York, NY.

Milman O. 2022. *The Insect Crisis: The Fall of the Tiny Empires That Run the World*. W. W. Norton & Company. New York, NY.

Cultural Entomology

Berenbaum MR. 1995. *The Bugs in the System: Insects and Their Impacts on Human Affairs*. Basic Books. New York, NY.

Berenbaum MR. 2000. *Buzzwords: A Scientist Muses on Sex, Bugs and Rock 'n' Roll*. Joseph Henry Press. Washington, D.C.

Berenbaum MR. 2009. *The Earwig's Tail: A Modern Bestiary of Multi-legged Legends*. Harvard University Press. Cambridge, MA.

Clausen LW. 1954. *Insect Fact and Folklore*. Macmillan. New York, NY.

Evans M, ed. 2023. SEISMA 03: Entomology. *SEISMA Magazine*. Parabola Press. Oxford, UK. seismamag.com/03-edition-entomology

Greenfield AB. 2005. *A Perfect Red*. Harper Perennial. New York, NY.

Insects & Human Culture, youtube.com/playlist?list=PLpwzp9TeQanNYmG6xm7I0gh7YGmnocPiN

Klein BA, Klein A. 2016. Insects Incorporated: Database of Cultural Entomology. *http://culturalentomology.org/*

Kritsky G, ed. 2024. *A Cultural History of Insects, vol. 1-6*. Bloomsbury Publishing. London, UK.

Kuper P. 2025. *INSECTOPIA: A Natural History*. W. W. Norton & Co. New York, NY.

Melillo E. 2020. *The Butterfly Effect: Insects and the Making of the Modern World*. Knopf. New York, NY.

Reaktion Books (London): Animal series has nine insect and two arachnid books and counting. Example: Sleigh C. 2003. *Ant*.

Sear D. Cultural Entomology Digest. *http://culturalentomology.com/ced/*

Waldbauer G. 2009. *Fireflies, Honey, and Silk*. University of California Press. Berkeley, CA.

Wiedermann RN, Fisher JR. 2021. *The Silken Thread: Five Insects and Their Impacts on Human History*. Oxford University Press. Oxford, UK.

NOTES

PREFACE

The oldest known depiction by a human of an insect:

Cavernes du Volp, cavernesduvolp.com/en/

Bégouën R, et al. 1996. Enlène (Montesquieu-Avantès, Ariège). Marie T. Thiault et Roy JB, coord.: L'art préhistorique des Pyrénées. Paris: Réunion des musées nationaux, pp. 182–192.

Bégouën H, Bégouën L. 1928. Découvertes nouvelles dans la Caverne des Trois Frères a Montesquieu-Avantès (Ariège). *Annual Review of Anthropology.* 33:358–364.

INTRODUCTION

for more than 73 percent of all known animal species:

Numbers or proportions of known (= scientifically identified) species come from the Global Biodiversity Information Facility, gbif.org/species/1

and flight gave insects the world:

Engel M. 2018. *Innumerable Insects: The Story of the Most Diverse and Myriad Animals on Earth.* Union Square & Co. New York, NY. p. 47.

one of every three bites we consume:

Jordan A, et al. 2021. Economic dependence and vulnerability of United States agricultural sector on insect-mediated pollination service. *Environmental Science & Technology.* 55:2243–2253. doi.org/10.1021/acs.est.0c04786

humans manage . . . at least twenty-two species of insect pollinators:

Osterman J, et al. 2021. Global trends in the number and diversity of managed pollinator species. *Agriculture, Ecosystems & Environment.* 322:107653. doi.org/10.1016/j.agee.2021.107653

We rely on . . . key ingredients; responsible for several billion dollars; a single cow produces:

Losey JE, Vaughan M. 2006. The economic value of ecological services provided by insects. *BioScience.* 56:311–323. doi.org/10.1641/0006-3568(2006)56 [311:TEVOES]2.0.CO;2

1.5 billion people lack basic sanitation; 419 million... defecate in the open:

World Health Organization: Sanitation, who.int/news-room/fact-sheets/detail/sanitation

gigatons of carbon:

Bar-On YM, et al. 2021. The biomass distribution on Earth. *Proceedings of the National Academy of Sciences.* 115:6506–6511. doi.org/10.1073/pnas.1711842115

That's forty-four million Statues of Liberty:

What Things Weigh, whatthingsweigh.com

rolls their ball in a straight line:

Dacke M, et al. 2003. Insect orientation to polarized moonlight: An African dung beetle uses the moonlit sky to make a swift exit after finding food. *Nature.* 424:33. doi.org/10.1038/424033a

Australian Dung Beetle Project:

National Museum of Australia: Dung Beetles in Australia, nma.gov.au/defining-moments/resources/dung-beetles-in-australia

the little things that run the world:

Wilson EO. 1987. The little things that run the world* (the importance and conservation of invertebrates). *Conservation Biology.* 1:344–346. doi.org/10.1111/j.1523-1739.1987.tb00055.x

to a good approximation, all species are insects!:

May RM. 1986. Biological diversity: How many species are there? *Nature.* 324:514–515. doi.org/10.1038/324514a0

about 60 percent of identified animal species are insects:

1,114,071 insects out of 1,845,185 animals.

ants on Earth may exceed twenty quadrillion:

Schultheiss P, et al. 2022. The abundance, biomass, and distribution of ants on Earth. *Proceedings of the National Academy of Sciences.* 119:e2201550119. doi.org/10.1073/pnas.2201550119

farmers armed with chicken-feather dusters:

Warning Signals from the Apple Valleys, ICIMOD 2001; youtube.com/watch?v=Qxs9civLaf8

PRODUCTS

blood of dead gladiators to cure epilepsy:

Bethge P. 2009. Europe's "medicinal cannibalism": The healing power of death. *Spiegel Online.*

SILK

Dislodged from a high tree branch by a curious crow:

There are many versions of the tale of Leizu's discovery of silk. Nowhere could I find an explanation for how something as strongly anchored as a silk cocoon could have fallen from the mulberry branch into Leizu's teacup. I took the liberty of insinuating a crow into the tale. Elizabeth Jean Younce kindly followed suit by drawing the corvid culprit.

innovation may have actually come from an adviser to the emperor:

Kuhn D. 1984. Tracing a Chinese legend: In search of the identity of the "first sericulturalist." *T'oung Pao.* 70:213–245.

silk fragments dating from circa 2750 BCE:
Barnard 1975, cited in Kuhn 1984. p. 218.

Stash silkworm eggs:
Hunt P. 2011. *Late Roman Silk: Smuggling and Espionage in the 6th Century CE.* archive.
    is/20130626180730/http:/traumwerk.stanford.edu/philolog/2011/08/byzantine_
    silk_smuggling_and_e.html#selection-755.0-755.26

monopoly of Byzantine silk:
Muthesius A. 2003. Silk in the medieval world. In: *The Cambridge History of Western
    Textiles.* Cambridge University Press. Cambridge, UK. p. 326.

Can Nü:
I've replaced older romanized forms of Chinese names with pinyin (Leizu instead of
    Lei-tsu), thanks to suggestions by Michael Nylan, historian of China at UC Berkeley.

Lady of the Nine Palaces:
Werner (1922) referred to Can Nü as "Concubine of the Nine Palaces," but "concubine"
    connotes a level too low in early China to be used in this context, according to
    historian Michael Nylan.

seventy-five million pounds . . . of silk are produced each year:
Food and Agricultural Studies: Crops and Livestock Products, fao.org/faostat/en/
    #data/QP

Four times as many caterpillars perish for our silk:
225 billion silkworms / 55.4 billion food animals. 2023 U.S. Animal Kill Clock,
    animalclock.org

an annual memorial service:
Su N. 2022. Insects in Japanese culture, and one that saved the country. *American
    Entomologist.* 68:52–28. doi.org/10.1093/ae/tmac010

the domesticated silk moth is also deficient when it comes to perceiving odors:
Bisch-Knaden S, et al. 2014. Anatomical and functional analysis of domestication effects
    on the olfactory system of the silkmoth *Bombyx mori. Proceedings of the Royal Society
    B* 281:20132582. doi.org/10.1098/rspb.2013.2582

The practice still entails culling great numbers of moths:
Insect Suffering from Silk, Shellac, Carmine, and Other Insect Products,
    reducing-suffering.org/insect-suffering-silk-shellac-carmine-insect-products/

silk bags were used to hold powder charges:
National Museum of American History, americanhistory.si.edu/blog/

the very act that sparked World War I could have been prevented:
The Guardian: Tests Prove That a Bulletproof Silk Vest Could Have Stopped the First
    World War, theguardian.com/artanddesign/2014/jul/29/bulletproof-silk-vest-
    prevent-first-world-war-royal-armouries

transforming an agrarian society to a military power
Su N. 2022. Insects in Japanese culture, and one that saved the country. *American
    Entomologist.* 68:52–58. doi.org/10.1093/ae/tmac010

British prisoners received unexpected gift packages:
Atlas Obscura: How Millions of Secret Silk Maps Helped POWs Escape Their Captors in WWII, atlasobscura.com/articles/how-millions-of-secret-silk-maps-helped-pows-escape-their-captors-in-wwii

silk cocoons... serve as ornament, rattle, or other use:
Peigler R. 1994. Non-sericultural uses of moth cocoons. Denver Museum of Natural History.

dedicated the rest of his professional life to astronomy:
NYPL Digital Collections, digitalcollections.nypl.org/collections/the-trouvelot-astronomical-drawings-atlas#/?tab=about

masterful depiction of a caddisfly larva:
For more caddisfly art by Anker Odum, see: Wiggins G. 2005. *Caddisflies: The Underwater Architects*. University of Toronto Press. Toronto, Canada.

a tradition among Japanese children:
NHK for School, www2.nhk.or.jp/school/movie/clip.cgi?das_id=D0005300469_00000

Darwin's bark spider . . . produces the toughest known silk:
Agnarsson I, et al. 2010. Bioprospecting finds the toughest biological material: Extraordinary silk from a giant riverine orb spider. *PLoS ONE*. 5:e11234. doi.org/10.1371%2Fjournal.pone.0011234

spider silk . . . is stronger than steel on a per-weight basis:
Phys.org: Spider Silk Is a Wonder of Nature, but It's Not Stronger than Steel, phys.org/news/2013-06-spider-silk-nature-stronger-steel.html

cobweb art:
Cassirer I. 1956. Paintings on cobwebs. *Natural History*. 65:202–220.

Tyrolean "tongue choir":
Classic FM: This Austrian "Tongue Choir" Is a Deeply Distressing Choral Experience, classicfm.com/music-news/videos/austrian-tyrolean-tongue-choir-moscow-nights/

Clever entrepreneurs:
The history of harvesting spider silk is beautifully chronicled in Eleanor Morgan's *Gossamer Days*, and Simon Peters's *Golden Spider Silk*.

parodied [Bon's work] in his *Gulliver's Travels*:
Swift, J. 1726. *Gulliver's Travels*. Harper. New York, NY. pp. 231–232.

extracting silk from a spider:
Published 1710, and several times later, including in 1748; archive.org/details/b30525743/page/n1/mode/2up
Image came from the following edition: de Termeyer RM. 1807. *Opuscoli Scientifici d'Entomologia di fiscia e d'Agricoltura dell'Abate*. Milan, Italy. Vol.1, plate VI.

hints at the spider as creator:
Peers S. 2012. *Golden Spider Silk*. V&A Publishing. London, UK.

fifteen thousand silk filaments:
New Scientist: Spider Silk Spun into Violin Strings, newscientist.com/article/dn21540-spider-silk-spun-into-violin-strings/

violin body made of a composite of spider silk:

Lifegate: This Violin Created by an Italian Designer Is Made From Spiders' Silk, lifegate.com/spider-silk-violin

Classic FM: These Violins Are Made From Spiders' Silk, classicfm.com/discover-music/instruments/violin/features/spider-silk/

"as close as we can get to the dreams of spiders":

Morgan E. 2010. A short history of spiders' silk spinning machines. *Antenna*. 34:9–12.

spider ranches:

Owlcation: Black Widow Spider Silk in World War I and II, owlcation.com/stem/The-patriotic-black-widow-spiders-of-World-War-II

biomedically useful "super lens":

Lin CB, et al. 2020. Optimal photonic nanojet beam shaping by mesoscale dielectric dome lens. *Journal of Applied Physics*. 127:243110. doi.org/10.1063/5.0007611

Spider-Man!:

Lee S, Ditko S. 1962. *Amazing Fantasy #15*.

abide by James Kakalios's calculations:

Kakalios J. 2005. *The Physics of Superheroes*. Penguin Group Inc. New York, NY.

mimics natural processes observed in spiders:

Wu Y, et al. 2017. Bioinspired supramolecular fibers drawn from a multiphase self-assembled hydrogel. *Proceedings of the National Academy of Sciences*. 114:8163–8168. doi.org/10.1073/pnas.1705380114

from a cow's udder and baby hamster's kidney:

Lazaris A, et al. 2002. Spider silk fibers spun from soluble recombinant silk produced in mammalian cells. *Science*. 295:472–476. doi.org/10.1126/science.1065780

1/70,000 spider:

The New York Times Magazine: Got Silk, nytimes.com/2002/06/16/magazine/got-silk.html

"The silk gland of spiders and the milk gland of goats":

Bedini SA. 2005. Along came a spider—spinning silk for cross-hairs. *The American Surveyor*. amerisurv.com/2005/04/30/along-came-a-spider-spinning-silk-for-cross-hairs

Other transgenic organisms:

Heidebrecht A, et al. 2017. doi.org/10.1002/adma.201404234

Andersson M, et al. 2017. doi.org/10.1038/nchembio.2269

Foong CP, et al. 2020. doi.org/10.1038/s42003-020-1099-6

bacteria to produce bee silk:

Launch: Tara Sutherland: Bio-synthetic Silk, launch.org/innovators/tara-sutherland

bagworm moth's silk:

Yoshioka T, et al. 2019. A study of the extraordinarily strong and tough silk produced by bagworms. *Nature Communications*. 10:1469. doi.org/10.1038/s41467-019-09350-3

carbon nanotubes and they will spin reinforced silk:

Lepore E, et al. 2017. Spider silk reinforced by graphene or carbon nanotubes. *2D Materials*. 4:031013. doi.org/10.1088/2053-1583/aa7cd3

Wang Q, et al. 2016. Feeding single-walled carbon nanotubes or graphene to silkworms for reinforced silk fibers. *Nano Letters*. 16:6695–6700. doi.org/10.1021/acs.nanolett.6b03597

WAX

Egyptians sealed . . . with beeswax:

Clark G. 1942. Bees in antiquity. *Antiquity*. 16:208–215. doi.org/10.1017/S0003598X00017701

Nefertiti:

Crane E. 1983. *The Archaeology of Beekeeping*. Cornell University Press. Ithaca, NY. p. 244.

Beekeeping . . . well established in Egypt:

Kritsky G. 2017. Beekeeping from antiquity through the Middle Ages. *Annual Review of Entomology*. 62:249–264. doi.org/10.1146/annurev-ento-031616-035115

"Archaeology of the invisible":

Nasa JL, et al. 2022. Archaeology of the invisible: The scent of Kha and Merit. *Journal of Archaeological Science*. 141:105577. doi.org/10.1016/j.jas.2022.105577

charred and fragile remnants of bees:

Mazar A. 2018. *The iron age apiary at Tel Rehov, Israel*. In: *Beekeeping in the Mediterranean: From Antiquity to the Present*. New Moudania, Greece. pp. 40–49.

Tutankhamun's funerary mask; the great solar temple Shesepibre; tomb of Pabasa:

Kritsky G. 2015. *The Tears of Re: Beekeeping in Ancient Egypt*. Oxford University Press. Oxford, UK. p. 107.

Kritsky G. 2917. Beekeeping from antiquity through the Middle Ages. *Annual Review of Entomology*. 62:249–264. doi.org/10.1146/annurev-ento-031616-035115

like honey robbers in Australia do today:

Crane E. 2001. *The Rock Art of Honey Hunters*. International Bee Research Association. Cardiff, UK. pp. 48–50.

sophisticated act of honey robbing:

Martínez MB, et al. 2021. The rock art shelter of Barranco Gómez (Castellote, Teruel), a new Levantine rock art site in the Guadalope nucleus. *Trabajos de Prehistoria*. 78:164–178. diposit.ub.edu/dspace/bitstream/2445/179126/1/712822.pdf

most famous of the honey-hunting cave paintings:

Kritsky G. 2020. Revisiting prehistoric honey hunting at Bicorp, Spain. *American Bee Journal*. 160:1147–1250.

Tracing of photograph and paintings:

Hernandez-Pacheco E. 1924. *Las pinturas prehistoricas de las Cuevas de la Araña (Valencia)*. Comision de Investigaciones Paleontologicas y Prehistoricas. 34 (Serie Prehistoria 28). Museo Nacional de Ciencias Naturales, Madrid, Spain. simurg.bibliotecas.csic.es/view/CSIC000087673

age of [cave art's] creation:

Dating caves can sometimes be more difficult than dating cave paintings. A recent example of a widely-circulated fossil discovery was used as evidence to establish the paleoge-ography of an entire region. The prized fossil "in plain sight" appeared to show the first regional example of one of the earliest animals, from the Ediacaran age (*Dickinsonia tenuis*), but was later revealed to be the decaying remnants of a modern giant honey bee (*Apis dorsata*) nest.

Meert JG, et al. 2023. Stinging News: "*Dickinsonia*" discovered in the Upper Vindhyan of India not worth the buzz. *Gondwana Research*. 117:1–7. doi.org/10.1016/j.gr.2023.01.003

Mud [nests] . . . prized as data-rich evidence:

Finch D, et al. 2020. 12,000-year-old Aboriginal rock art from the Kimberley region, Western Australia. *Science Advances*. 6:eaay3922. doi.org/10.1126/sciadv.aay3922

objects of magic and sorcery:

May SK, et al. 2017. The rock art of Ingaanjalwurr, western Arnhem Land, Australia. In: *The Archaeology of Rock Art in Western Arnhem Land, Australia*. ANU Press, Canberra, Australia.

loud trill that ends with a short grunt:

Listen to the signals made by humans to communicate with honeyguides: youtube.com/watch?v=hGC4nG0RqYI

increase your chances of engaging a willing honeyguide:

Spottiswoode CN, et al. 2022. Honeyguides. *Current Biology*. 32:R1072–R1074.

Spottiswoode CN, et al. 2016. Reciprocal signaling in honeyguide-human mutualism. *Science*. 353:387–389. doi.org/10.1126/science.aaf4885

"far more lifelike and lovelier":

For beautiful writings on morbid phenomena by Joanna Ebenstein see: Ebenstein J, Dickey C. 2014. *The Morbid Anatomy Anthology*. Morbid Anatomy Press. Brooklyn, NY. p. 70.

vestiges of insect anatomy:

Pollett DK. 1972. *The Morphology, Biology and Control of* Ceroplastes ceriferus *(Fabricius) and* Ceroplastes sinensis Del Guercio *in Virginia Including a Redescription of* Ceroplastes floridensis Comstock *(Homoptera: Coccoidea: Coccidae)*. Virginia Tech. Blacksburg, VA. p. 79.

Forensic entomologists are attempting to use these same chemicals:

Sharma A, et al. 2021. Cuticular hydrocarbons as a tool for determining the age of *Chrysomya rufifacies* (Diptera: Calliphoridae) larvae. *Journal of Forensic Sciences*. 66:236–244. doi.org/10.1111/1556-4029.14572

DNA trapped within louse cement:

Pedersen MW, et al. 2021. Ancient human genomes and environmental DNA from the cement attaching 2,000-year-old head lice nits. *Molecular Biology and Evolution*. 39:msab351. doi.org/10.1093/molbev/msab351

HONEY

Enticed with gifts and invited ashore:

Robert Knox: Ancient Ceylon's Most Famous British Captive, ceylon.guide/2020/04/24/
robert-knox-ancient-ceylons-most-famous-british-captive/

"put honey in it, and then fill it up with flesh":

Knox R. 1681. (1911). *A Historical Relation of Ceylon*. J. MacLehose & Sons. Glasgow, UK.

buried a pantry full of foods and weavings coated in honey:

National Geographic: Out of Eden Walk: Honey I'm Dead, nationalgeographic.org/
projects/out-of-eden-walk/articles/2015-05-honey-im-dead/

embalmed with honey:

Anthropologist Ajit K. Danda, when asked about preserving flesh with honey, shared
with me lessons from Indian colleagues: "Among the Khasia of Meghalaya State
there was a general notion in circulation about use of honey for preservation of raw
flesh . . . Although the local people shared the knowledge of such a practice, they do
not use honey any more to preserve raw flesh. Instead, they use sunlight and smoke
for such preservation. But use of honey as the pain-killer is still found in practice
. . . Tribes of Andaman and Nicobar islands have multiple uses of honey but not
to preserve raw flesh. The pregnant women there usually receive a special treat of
extra-consumption of honey, for the protection of health of the expectant mother as
well as of the baby in the womb."

Alexander the Great . . . may have been submerged in honey-filled caskets:

Crane E. 1983. *The Archaeology of Beekeeping*. Cornell University Press. Ithaca, NY. p. 240.

Earls of Southampton, immersed in honey:

Vernon F. 1981. *Hogs at the Honeypot*. Bee Books New & Old. Steventon, UK.

"for the consideration of the learned":

Needham J, Lu G. 1974. *Science and Civilization in China*. Vol. 5, Part II. Cambridge
University Press. Cambridge, UK. p. 76.

**"All bees in making honey carry human feces to the flowers"; "one becomes a divine
immortal":**

Read BE. 1982. *Chinese Materia Medica: Insect Drugs, Dragon & Snake Drugs, Fish Drugs*.
Southern Materials Center, Inc. pp. 12, 15.

**honey contains roughly two hundred substances; Hippocrates also prescribed
honey for contraception; honey can inhibit about sixty species of bacteria:**

Eteraf-Oskouei T, Najafi M. 2013. Traditional and modern uses of natural honey in
human diseases: A review. *Iranian Journal of Basic Medical Sciences*. 16:731–742.

**about five hundred [treatments for ailments] include honey; honey . . . and croco-
dile dung:**

Kritsky G. 2015. *The Tears of Re: Beekeeping in Ancient Egypt*. Oxford University Press.
Oxford, UK. pp. 91–92.

Honey may also serve as a future indicator of air quality:

Smith KE, et al. 2019. Honey as a biomonitor for a changing world. *Nature Sustainability*.
2:223–232. doi.org/10.1038/s41893-019-0243-0

fueled their ever-expanding brains?:
Crittenden AN. 2011. The importance of honey consumption in human evolution. *Food and Foodways*. 19:257–273. doi.org/10.1080/07409710.2011.630618

LACQUER
Lacquer is a liquid or paste that dries to form a semitransparent protective or decorative coating. Most lacquers are derived from plants, but there are insect exceptions.

COLOR
if any non-Caesar were caught wearing a purple garment:
The University of Chicago Library: Tyrian Purple, lib.uchicago.edu/collex/exhibits/originsof-color/organic-dyes-and-lakes/tyrian-purple/
seeing red can bias our attention:
Elliot AJ. 2015. Color and psychological functioning: A review of theoretical and empirical work. *Frontiers in Psychology*. 6:368. doi.org/10.3389/fpsyg.2015.00368
how sexually attractive we find someone:
Psychology Today: Red Alert, psychologytoday.com/us/blog/insight-therapy/201301/red-alert-science-discovers-the-color-sexual-attraction
The Atlantic: Seeing Red, theatlantic.com/health/archive/2014/12/seeing-red/383048/
respond faster and with greater force:
Elliot AJ, Aarts H. 2011. Perception of the color red enhances the force and velocity of motor output. *Emotion*. 11:445–449. doi.org/10.1037/a0022599
red triumphs over blue:
Hill R, Barton R. 2005. Red enhances human performance in contests. *Nature*. 435:293. doi.org/10.1038/435293a
the oil contains cosmetic pigments:
From So Simple a Beginning: Flamingos Use Cosmetics, blog.willyvanstrien.nl/2018/07/29/flamingos-use-cosmetics
ten thousand snails killed to produce one gram:
World History Encyclopedia: Tyrian Purple, worldhistory.org/Tyrian_Purple
wormberry:
For dramatic tales of our history with the color red: Greenfield AB. 2005. *A Perfect Red*. Harper Perennial. New York, NY. p.126.
Two organisms not deterred [by carminic acid]
Eisner T, et al. 1980. Red cochineal dye (carminic acid): Its role in nature. *Science*. 208:1039–1042. doi.org/10.1126/science.208.4447.1039
to produce different colors:
Andean dyers selected cochineal bugs for their larger bodies (containing more carminic acid), developed techniques of adding a mordant to help bond carmine to silk and mammal hairs, and used other additives for a range of hues.
Dactylopius coccus:
Dactylopius confusus was also cultivated in Peru.

may be the only living person who can create the red ink:
Personal communication with Yavuz Parlar (studio and production manager), and
    reported on Çavuşoğlu's website, aslicavusoglu.info/2015/red-red

all three versions of *The Bedroom*:
Berns RS. 2019. Digital color reconstructions of cultural heritage using color-managed
    imaging and small-aperture spectrophotometry. *Color Research and Application.*
    44:531–546. doi.org/10.1002/col.22371

oak gall ink:
You knew that *The Iron Gall Ink Website* must exist. Here is where you can find your
    favorite recipes: irongallink.org/index.html

Magna Carta:
UK Parliament: The Making of Magna Carta, parliament.uk/about/living-heritage/
    evolutionofparliament/2015-parliament-in-the-making/2015-historic-anniversaries/
    magna-carta/the-making-of-magna-carta

increasing our heart rate:
AL-Ayash A, et al. 2015. The influence of color on student emotion, heart rate, and
    performance in learning environments. *Color Research and Application.* 41:196–205.
    doi.org/10.1002/col.21949

red has its advantages:
ScienceDaily: Color Red Increases the Speed and Strength of Reactions, sciencedaily.com/
    releases/2011/06/110602122349.htm

"vegan" options exist:
An Irish luthier has made violins registered with The Vegan Society's Vegan Trademark,
    veganviolins.com

The bulk of a violin is made from woody plants:
Brian Derber, author of the most beautiful, comprehensive, and viable manual to creating
    a violin, refers to cochineal, kermes, propolis, shellac, and other insect products.
Derber BT. 2017. *The Manual of Violin Making: A Supplemental Textbook of a School of Violin
    Making in the Modern Age.* Brian T. Derber, Presque Isle, Wisconsin. pp. 311, 314–315.

Sources of ivory include teeth and tusks:
I did not expect to find "fossil mammoth ivory" for sale through Guitar Parts and More,
    but the Warther family of Dover, Ohio, carves ancient bone into bow frogs and buttons.
guitarpartsandmore.com/productCategory.php?Mammoth-Ivory-Violin-Bow-Frog-
    and-Button-Blanks-104

violins built by Antonio Stradivari:
The Strad: What Is the Secret Behind Stradivari's Red Violins? thestrad.com/lutherie/
    what-is-the-secret-behind-stradivaris-red-violins/8128.article

as my sister astutely points out:
Korinthia Klein, musician, luthier, and author of several novels, wrote a guide for violin
    teachers, players, or parents of players to better understand their instruments. She
    has managed to write a manual that is fun to read even if you are none of the above.
Klein KA. 2020. *My Violin Needs Help! A Diagnostics Guide for Players and Teachers.*
    Self-published.

PAPER

"The rags from which we make our paper":

Réaumur, from Hunter D. 1967. *Papermaking: The History and Technique of an Ancient Craft.* 2nd ed. Knopf. New York, NY. (translated from the French) pp. 314–315.

CHITIN

Traversing the cosmic void poses challenges:

NASA: 5 Hazards of Human Spaceflight, nasa.gov/hrp/5-hazards-of-human-spaceflight

Chitin . . . has attractive, eco-friendly properties:

Machałowski T, et al. 2020. Chitin of Araneae origin: Structural features and biomimetic applications: a review. *Applied Physics A.* 126:678. doi.org/10.1007/s00339-020-03867-x

eco-friendly extraction methods:

Mohan K, et al. 2022. Green and eco-friendly approaches for the extraction of chitin and chitosan: A review. *Carbohydrate Polymers.* 287:119349. doi.org/10.1016/j.carbpol.2022.119349

chitin or chitosan has potential applications:

Gortari MC, Hours RA. 2013. Biotechnological processes for chitin recovery out of crustacean waste: A mini-review. *Electronic Journal of Biotechnology.* 16. doi.org/10.2225/vol16-issue3-fulltext-10

surgical thread and sutures:

Nakajima M, et al. 1986. Chitin is an effective material for sutures. *The Japanese Journal of Surgery.* 16:418–424. doi.org/10.1007/BF02470609

scaffolds for tissue engineering:

Machałowski T, et al. 2020. Chitin of Araneae origin: Structural features and biomimetic applications: A review. *Applied Physics A.* 126:678. doi.org/10.1007/s00339-020-03867-x

building material that could help support humans on Mars:

Shiwei N, et al. 2020. Martian biolith: A bioinspired regolith composite for closed-loop extraterrestrial manufacturing. *PLoS ONE.* 15:e0238606. doi.org/10.1371/journal.pone.0238606

fruit flies and plant seeds boarded German V-2 rockets:

Beischer DE, Fregly AR. 1962. Animals and man in space. A chronology and annotated bibliography through the year 1960. US Naval School of Aviation Medicine. apps.dtic.mil/sti/tr/pdf/AD0272581.pdf

VENOM AND POISON

"There are gods, and they do throw thunderbolts":

Schmidt JO. 2016. *The Sting of the Wild: The Story of the Man Who Got Stung for Science.* Johns Hopkins University Press. Baltimore, MD.

lacing their weapons with venoms from scorpions:

Steyn (1949) published the first major list of about three hundred plants and fifteen animals used as arrow poisons in central to southern Africa; cited by: Chaboo C, et al. 2019. Beetle and plant arrow poisons of the San people of southern Africa. In: *Toxicology in Antiquity*, 2nd ed. Academic Press.

Wikar in his 1779 journal:

Attempting to remove offensive terms at the expense of altering direct quotes, I replaced "The Hottentots" with "[People]." pp. 179–181.

larvae of different species:

For an overview of the beetle poisons used by the San of southern Africa, and references for further research on the fascinating topic see Chaboo CS, et al. 2016. Beetle and plant arrow poisons of the Ju|'hoan and Hai||om San peoples of Namibia (Insecta, Coleoptera, Chrysomelidae; Plantae, Anacardiaceae, Apocynaceae, Burseraceae). *ZooKeys.* 558: 9–54. doi.org/10.3897/zookeys.558.5957

unchanged in their toxicity when tested eighty-eight years later:

Shaw EM, et al. 1963. *Bushman arrow poisons.* Cimbebasia. Windhoek, Namibia. 7:2–41. p. 11.

promising tests exist with venoms of ants:

Pachycondyla sennaarensis

Al-Tamimi J, et al. 2021. Samsum ant venom protects against carbon tetrachloride-induced acute spleen toxicity in vivo. *Environmental Science and Pollution Research.* 28:31138–31150. doi.org/10.1007/s11356-020-12252-3

Solenopsis invicta

Roy A., Bharadvaja N. 2020. Venom-derived bioactive compounds as potential anticancer agents: A review. *International Journal of Peptide Research and Therapeutics.* 27:129–147. doi.org/10.1007/s10989-020-10073-z

and wasps:

Vespa velutina

Yun HS, et al. 2021. Anti-inflammatory effect of wasp venom in BV-2 microglial cells in comparison with bee venom. *Insects.* 29:297. doi.org/10.3390/insects12040297

V. magnifica

Gao Y, et al. 2020. Wasp venom possesses potential therapeutic effect in experimental models of rheumatoid arthritis. *Evidence-Based Complementary and Alternative Medicine.* 2020:6394625. *doi.org/10.1155/2020/6394625*

cancer-fighting activity:

Suttmann H, et al. 2008. Antimicrobial peptides of cecropin family show potent antitumor activity against bladder cancer cells. *BMC Urol.* 8:5. doi.org/10.1186/1471-2490-8-5

Pal P, et al. 2015. Medicinal value of animal venom for treatment of cancer in humans—a review. *International Journal of Peptide Research and Therapeutics.* 22:128–144.

funnel-web spider can help with cardiovascular disease:

Redd MA, et al. 2021. Therapeutic inhibition of acid-sensing ion channel 1a recovers heart function after ischemia-reperfusion injury. *Circulation.* 144:947–960. doi.org/10.1161/CIRCULATIONAHA.121.054360

deathstalker scorpion can uncover . . . tumors:
Soroceanu L, et al. 1998. Use of chlorotoxin for targeting of primary brain tumors. *Cancer Research.* 58:4871–4879.
The New York Times Magazine: Deadly Venom from Spiders and Snakes May Also Cure What Ails You, nytimes.com/2022/05/03/science/venom-medicines.html
chimpanzee troop . . . using some mystery insect for possible self-medication:
As I write this, the authors will be closer to identifying the mystery insect!
Mascaro A, et al. 2022. Application of insects to wounds of self and others by chimpanzees in the wild. *Current Biology.* 32:PR112–R113. doi.org/10.1016/j.cub.2021.12.045
the agony and temporary paralysis of dozens of ants injecting their venom:
Bosmia AN, et al. 2015. Ritualistic envenomation by bullet ants among the Sateré-Mawé Indians in the Brazilian Amazon. *Wilderness Environmental Medicine.* 26:271–273. doi.org/10.1016/j.wem.2014.09.003
caterpillar of a moth, possibly *Myelobia (Morpheis) smerintha*:
Britton EB. 1984. A pointer to a new hallucinogen of insect origin. *Journal of Ethnopharmacology.* 12:331–333. doi.org/10.1016/0378-8741(84)90061-8
"When I was among the Malalis":
Quote by Augustin de Saint-Hilaire (1824), translated from the French and republished. Jenkins A (Ed.). 1946. *Chronica Botanica.* 10:24–61.
used the ants to acquire supernatural, shamanic powers:
For vivid descriptions of the practice, read Peter Groark's ethnographic account.
Groark K. 1996. Ritual and therapeutic use of "hallucinogenic" harvester ants (*Pogonomyrmex*) in native South-Central California. *Journal of Ethnobiology* 16:1–29.
despite the risk of suffering from food poisoning or worse:
Koca I, Koca AF. 2007. Poisoning by mad honey: A brief review. *Food and Chemical* 45:1315–1318. doi.org/10.1016/j.fct.2007.04.006
mad honey . . . to increase their sexual arousal:
Demircan A, et al. 2009. Mad honey sex: Therapeutic misadventures from an ancient biological weapon. *Annals of Emergency Medicine.* 54:824–829. doi.org/10.1016/j.annemergmed.2009.06.010
using animals without backbones for medicinal reasons:
Meyer-Rochow VB. 2017. Therapeutic arthropods and other, largely terrestrial, folk-medicinally important invertebrates: A comparative survey and review. *Journal of Ethnobiology and Ethnomedicine.* 13:9. doi.org/10.1186/s13002-017-0136-0
"ant Viagra":
Pemberton RW. 2023. "Ant Viagra" in the Cameroon. American Entomologist. 69:34–37. doi.org/10.1093/ae/tmad019
***Chinese Materia Medica*:**
Read BE. 1982. *Chinese Materia Medica: Insect Drugs, Dragon & Snake Drugs, Fish Drugs.* Southern Materials Center, Inc. pp. 66 (stink bug), 75 (cicada), 73 (dragonflies).

gross deformities in a variety of animals:

Zainullin VG, et al. 1992. The mutation frequency of *Drosophila melanogaster* populations living under conditions of increased background radiation due to the Chernobyl accident. *Science of the Total Environment*. 112:37–44. doi.org/10.1016/0048-9697(92)90236-l

Møller AP. 2002. Developmental instability and sexual selection in stag beetles from Chernobyl and a control area. *Ethology*. 108:193–204. doi.org/10.1046/j.1439-0310.2002.00758.x

Medvedev LN. 1996. Chrysomelidae and radiation. In: *Chrysomelidae Biology*. SPB Academic Publishing. Amsterdam, Netherlands. pp. 403–410.

severe abnormalities following the Fukushima Daiichi Nuclear Power Plant catastrophe:

Hiyama A, et al. 2012. The biological impacts of the Fukushima nuclear accident on the pale grass blue butterfly. *Scientific Reports*. 2:570. doi.org/10.1038/srep00570

Even functional nuclear power plants appear connected to insect aberrations:

Hesse-Honegger C, Wallimann P. 2008. Malformation of true bug (Heteroptera): A phenotype field study on the possible influence of artificial low-level radioactivity. *Chemistry and Biodiversity*. 5:499–539. doi.org/10.1002/cbdv.200800001

Körblein A, Hesse-Honegger C. 2018. Morphological abnormalities in true bugs (Heteroptera) near Swiss nuclear power stations. *Chemistry and Biodiversity*. 15:e1800099. doi.org/10.1002/cbdv.201800099

An asymmetrical shield bug:

Hesse-Honegger C. 1995. *The Future's Mirror*. Locus+. Newcastle upon Tyne, UK.

grasshoppers . . . respond to the smell of bombs:

Saha D, et al. 2020. Explosive sensing with insect-based biorobots. *Biosensors and Bioelectronics: X*. 6:100050. doi.org/10.1016/j.biosx.2020.100050

Ants can be trained to detect cancer cells:

Piqueret B, et al. 2022. Ants detect cancer cells through volatile organic compounds. *iScience*. 22:103959. doi.org/10.1016/j.isci.2022.103959

employing bees to detect landmines:

Hadagali MD, Suan CL. Advancement of sensitive sniffer bee technology. *TrAC Trends in Analytical Chemistry*. 97:153–158. doi.org/10.1016/j.trac.2017.09.006

and COVID-19:

The Washington Post: Scientists May Have Found a New Coronavirus Rapid-Testing Method: Bees, washingtonpost.com/science/2021/05/07/covid-bee-testing

bees act as crop dusters:

Popular Science: Meet the Tiniest Crop Duster, popsci.com/meet-tiniest-crop-duster

ward away African elephants:

ScienceShot: "Beehive Fences" Keep Elephants Out, science.org/content/article/scienceshot-beehive-fences-keep-elephants-out

Feed black soldier flies . . . and process biofuels:

Qing L, et al. 2011. Bioconversion of dairy manure by black soldier fly (Diptera: Stratiomyidae) for biodiesel and sugar production. *Waste Management*. 31:1316–1320. doi.org/10.1016/j.wasman.2011.01.005

can be eaten by waxworms . . . and mealworms:

Bombelli P, et al. 2017. Polyethylene bio-degradation by caterpillars of the wax moth *Galleria mellonella*. *Current Biology*. 27:R292–R293. doi.org/10.1016/j.cub.2017.02.060

Yang Y, et al. 2015. Biodegradation and mineralization of polystyrene by plastic-eating mealworms: Part 1. Chemical and physical characterization and isotopic tests. *Environmental Science and Technology*. 49:12080–12086. doi.org/10.1021/acs.est.5b02661

exploited insects in gruesome ways during times of war:

Lockwood JA. 2008. *Six-Legged Soldiers: Using Insects as Weapons of War*. Oxford University Press. Oxford, UK.

fears of terrorists using insects:

A tale I once read related a scene in which insects were used to thwart suffering caused by war, rather than perpetuate it. As the army of Kublai Khan (grandson of Genghis Khan) approached, villagers feverishly painted sugar water on plants. Caterpillars ate the sweet brush marks, leaving a warning on every leaf of every tree: "It is the will of Heaven. The invaders must leave." Keith Taylor, Sino-Vietnamese Cultural Studies professor at Cornell University, wrote to me that this tale was invented or imagined "amidst the exuberance of modern Vietnamese nationalism. . . . There is historical evidence," Taylor added, "that at least some Vietnamese soldiers were tattooed, perhaps on their foreheads, with the words 'Death to the Invaders.'"

FOOD

"Hemiptera crews":

Poet, playwright, and novelist, Bill Harris wrote a collection of insect poems to accompany Karen Anne Klein's art in the hand-made artist book, *Tiny Beasts* (2013, edition of 22). In "WASP's soliloquy," a wasp laments the undeserved, bad reputation shared by all wasps: "And so on they drone. Remarks so out of joint they'd cut us, if we had them, to the very bone . . ."

only scientifically described in 2013:

Hajong SR, Yaakop S. 2013. *Chremistica ribhoi* sp. n. (Hemiptera: Cicadidae) from North-East India and its mass emergence. *Zootaxa*. 3702:493–500. doi.org/10.11646/zootaxa.3702.5.8

But why the thirteen- or seventeen-year wait?:

Williams K, Simon C. 1995. The ecology, behavior, and evolution of periodical cicadas. *Annual Review of Entomology*. 40:269–295. doi.org/10.1146/annurev.en.40.010195.001413

Reverend Andreas Sandel wrote in his journal:

Kritsky G. 2021. One for the books: The 2021 emergence of the periodical cicada brood X. *American Entomologist*. 67:40–46. doi.org/10.1093/ae/tmab059

1,611 species of insects have traditionally been consumed:

Itterbeek JV, Pelozuelo L. 2022. How many edible insect species are there? A not so simple question. *Diversity.* 14:143. doi.org/10.3390/d14020143

DNA from a single tea bag:

Krehenwinkel H, et al. 2022. The bug in a teacup—monitoring arthropod–plant associations with environmental DNA from dried plant material. *Biology Letters.* 18:2022009120220091 doi.org/10.1098/rsbl.2022.0091

used a small subset of these bones to dig for termites:

Backwell LR, d'Errico F. 2001. Evidence of termite foraging by Swartkrans early hominids. *Proceedings of the National Academy of Sciences.* 98:1358–1363. doi.org/10.1073/pnas.98.4.1358

patterns of wear on some ancient bones:

Lesnik JJ. 2018. *Edible Insects and Human Evolution.* University Press of Florida, Gainesville, FL.

markings indicative of termite mound excavation:

Hanon R, et al. 2021. New evidence of bone tool use by Early Pleistocene hominins from Cooper's D, Bloubank Valley, South Africa. *Journal of Archaeological Science: Reports.* 39:103129. doi.org/10.1016/j.jasrep.2021.103129

Bone tools from the early hominid sites:

Backwell LR, d'Errico F. 2008. Early hominid bone tools from Drimolen, South Africa. *Journal of Archaeological Science.* 35:2880–2894. doi.org/10.1016/j.jas.2008.05.017

spread of human colonialism:

Lesnik JJ. 2019. The colonial/imperial history of insect food avoidance in the United States. *Annals of the Entomological Society of America.* 112:560–565. doi.org/10.1093/aesa/saz023

A Western-driven stigma against eating insects:

DeFoliart GR. 1999. Insects as food: Why the Western attitude is important. *Annals of the Entomological Society of America.* 44:21–50. doi.org/10.1146/annurev.ento.44.1.21

Weaver ants . . . can themselves be harvested as a sustainable food source:

Offenberg J, Wiwatwitaya D. 2009. Weaver ants convert pest insects into food—prospects for the rural poor. *Conference: Tropentag 2009, University of Hamburg.* Hamburg, Germany.

Insect welfare and the ethics of consuming insects:

Delvendahl N, et al. 2022. Edible insects as food—insect welfare and ethical aspects from a consumer. *Insects.* 13:121. doi.org/10.3390/insects13020121

lack of familiarity with insects among the translators [of the King James Version]:

Isman MB, Cohen MS. 1995. Kosher insects. *American Entomologist.* 41:100–103. doi.org/10.1093/ae/41.2.100

Readers of the Koran will find mention of eight insects:

Sira Juddin AS, et al. 2021. Exegetes' interpretations on insects mentioned in the Quran: Comparison between Tafsir Ibn Kathir and Tafsir Al-Azhar. *Proceedings of the 7th International Conference on Quran as Foundation of Civilization* (SWAT 2021). pp. 63–75. oarep.usim.edu.my/jspui/handle/123456789/14243

farming insects emits a lower amount of greenhouse gases per calorie:
Oxford Martin School. 2019. *Meat: The Future Series*. World Economic Forum. Geneva,
Switzerland. weforum.org/whitepapers/meat-the-future-series-alternative-proteins

champion of entomophagy, Arnold van Huis:
van Huis A. 2013. Potential of insects as food and feed in assuring food security. *Annual
Review of Entomology*. 58:563–583. doi.org/10.1146/annurev-ento-120811-153704

how best to shift . . . to insect-based food:
Berggren A, et al. 2022. Approaching ecological sustainability in the emerging insects-
as-food industry. *Trends in Ecology & Evolution*. 34:132–138. doi.org/10.1016/
j.tree.2018.11.005

Delvendahl N, et al. 2022. Edible insects as food—insect welfare and ethical aspects from
a consumer perspective. *Insects*. 13:121. doi.org/10.3390/insects13020121

it tasted remarkably like blue cheese:
Penick CA, Smith AA. 2015. The true odor of the odorous house ant. *American Entomologist*. 61:85–87. doi.org/10.1093/ae/tmv023

Wired: The Ants That Smell Like Blue Cheese—Or Is That Pine-Sol? wired.com/2015/
06/ants-smell-like-blue-cheese-pine-sol

MEDICINE

"made into pills with human blood from a man slain in battle":
Read BE. 1982. *Chinese Materia Medica: Insect Drugs, Dragon & Snake Drugs, Fish Drugs*.
Southern Materials Center, Inc. pp. 32–34.

discovering drug resources from nature:
Costa-Neto EM. 2002. The use of insects in folk medicine in the state of Bahia, North-
eastern Brazil, with notes on insects reported elsewhere in Brazilian folk medicine.
Human Ecology. 30:245–263. doi.org/10.1023/A:1015696830997

how many insects have been used medicinally:
Meyer-Rochow VB. 2017. Therapeutic arthropods and other, largely terrestrial, folk-
medicinally important invertebrates: A comparative survey and review. *Journal of
Ethnobiology and Ethnomedicine*. 13:2. doi.org/10.1186/s13002-017-0136-0

used in drug delivery systems:
Park BK, Kim M. 2010. Applications of chitin and its derivatives in biological medicine.
International Journal of Molecular Sciences. 11:5152–5164. doi.org/10.3390/
ijms11125152

polyethylene glycol . . . can cause anaphylaxis:
Sellaturay P, et al. 2021. Polyethylene glycol (PEG) is a cause of anaphylaxis to the Pfizer/
BioNTech mRNA COVID-19 vaccine. *Clinical and Experimental Allergy*. 51:861–863.
doi.org/10.1111/cea.13874

[Novavax] includes proteins created by cells from moths:
Nebraska Medicine: Moths and Tree Bark: How the Novavax Vaccine Works,
nebraskamed.com/COVID/moths-and-tree-bark-how-the-novavax-
vaccine-works

formidable jaws . . . to close wounds:

Kutalek R. 2011. Ethnoentomology: A neglected theme in ethnopharmacology? *Journal of Medical Anthropology*. 34:128–136.

"The barber presses together the lips of the wound":

The quote (an excerpt, originally in French) comes from Marcel Baudouin (1898). Entomologist Ulrich Mueller sent me a photocopy of this article with a Post-it Note attached: "No need for health insurance, really."

Gudge, EW. 1925. Stitching wounds with the mandibles of ants and beetles: A minor contribution to the history of surgery. *Journal of the American Medical Association*. 84:1861–1864.

honey bee's bite (not sting) comes with a local anesthetic:

Papachristoforou A, et al. 2012. The bite of the honeybee: 2-heptanone secreted from honeybee mandibles during a bite acts as a local anesthetic in insects and mammals. *PLoS ONE*. 7:e47432. doi.org/10.1371/journal.pone.0047432

LIGHT

the more *Photinus* a *Photuris* female consumes:

Eisner T, et al. 1997. Firefly "femmes fatales" acquire defensive steroids (lucibufagins) from their firefly prey. *Proceedings of the National Academy of Sciences*. 94:9723–9728. doi.org/10.1073/pnas.94.18.9723

Fireflies' nocturnal glow is depicted in countless pieces of art; art with fireflies:

Prischmann-Voldseth, DA. 2022. Fireflies in art: Emphasis on Japanese woodblock prints from the Edo, Meiji, and Taishō periods. *Insects*. 13:775. doi.org/10.3390/insects13090775

Neanderthals . . . lightning beetle biochemistry:

The New York Times Magazine: New DNA Test is Yielding Clues to Neanderthals, nytimes.com/2006/11/16/science/16neanderthal.html

tying these beetles to their toes:

Hogue CL. 1993. *Latin American Insects and Entomology*. University of California Press. Berkeley, CA. p. 259.

An added bonus: The explorer and naturalist Alexander von Humboldt noted that placing a dozen headlight beetles into a perforated gourd served well as a reading lamp.

ART

When Yanagi prepares a work of ant art:

The earliest versions of Yanagi's pieces were backed by Uncle Milton Ant Farms, a novelty created by Milton Levine in 1956 that still sells today.

insect media art:

Klein BA. 2022. Wax, wings, and swarms: Insects and their products as art media. *Annual Review of Entomology*. 67:281–303. doi.org/10.1146/annurev-ento-020821-060803

A grasshopper became mired in oil paint on one of Vincent van Gogh's canvases:
An insect in the paint was not a unique event. Van Gogh wrote, in a letter to his brother
Theo (ca. 14 July 1885) "I must have picked a good hundred flies and more off the
4 canvases that you'll be getting, not to mention dust and sand &c." Translated by
The Van Gogh Letters Project team. vangoghletters.org/vg/letters/let515/letter.html

"singing shawls":
Rivers, VZ. 2003. Emeralds on wing: Jewel beetles in textiles and adornment. In: *Insects
in Oral Literature and Traditions*. Peeters. Paris, France. pp. 163–175.

Museum of Jurassic Technology:
The museum forces visitors to question what is real and what is fantasy. Plan to visit, and
to read Lawrence Weschler's *Mr. Wilson's Cabinet of Wonders*. mjt.org/exhibits/
dalton/dalton.html

miniature, staged dramas made with modified train set figures:
Slinkachu describes the process on his website: "I often remodel the characters, cutting
them up and reposing them, adding new features such as hoods with modelling clay,
or changing arm and leg positions. I then paint the characters with acrylic paint and,
if needed, find props for them."

the earliest stop-motion puppetry in film history:
Ladislas Starewitch's granddaughter, Léona Béatrice Martin-Starewitch, and her husband,
François Martin, manage Starewitch's legacy. Martin-Starewitch wrote to let me
know that works of his still exist in their archives: "There are puppets, sets, his personal insect and butterfly collection . . ." starewitch.fr

In Fairyland:
Enchanting quotes abound in this book about Tessa Farmer's art (Catriona McAra: p. vii;
Giovanni Aloi: p. 94; Gail-Nina Anderson: p. 48; Petra Lange-Berndt: pp. 79, 75;
Gavin Broad: p. 29).

These tiny fairies wield weapons consisting of . . . wasp stingers:
This is not the only time insect stingers have played a menacing role in art. As I reported
in the *Annual Review of Entomology*, artist Judith Klausner constructed a crown of
thorns composed of honey bee stingers in *Apis Ignota, Operaria Alvi* (2009). Klausner described the process to me: "I extracted stingers from dead honey bees, which
was tougher than I thought it would be. I had to crush the abdomen and root around
in the wreckage for a tiny sliver and extract it with forceps. I used the stingers as the
thorns in the crown of thorns, attached around a circlet made from one of my hairs."

SECTION II

"and the lifespan of a mayfly!":
Mayflies (order Ephemeroptera, derived from the Greek words meaning "lasting a day—
wing") can live extremely short lives as adults. The (short) record holder is a female
Dolania americana that lives for a mere five minutes as an adult. To be successful, she
needs to molt, mate, and lay eggs without being eaten by predators, all within three
hundred seconds or less.

Sweeney BW, Vannote RL. 1982. Population synchrony in mayflies: A predator satiation hypothesis. *Evolution.* 36:810–821. doi.org/10.2307/2407894

evolution acts more like a tinkerer:

François Jacob famously thought about natural selection as a tinkerer. "Evolution behaves like a tinkerer who, during eons upon eons, would slowly modify his work, unceasingly retouching it, cutting here, lengthening there, seizing the opportunities to adapt it progressively to its new use."

Jacob F. 1977. Evolution and tinkering. *Science.* 196:1161–1166. doi.org/10.1126/science.860134

only 1 percent of all species to have ever evolved on Earth:

Barnosky AD, et al. 2011. Has the Earth's sixth mass extinction already arrived? *Nature.* 471:51–57. doi.org/10.1038/nature09678

Our World in Data: Biodiversity. ourworldindata.org/extinctions#how-many-species-have-gone-extinct

glue on feather extensions:

Andersson M. 1982. Female choice selects for extreme tail length in a widowbird. *Nature.* 299: 818–820. doi.org/10.1038/299818a0

smallest insect ever described is a male "fairyfly":

Mockford EL. 1997. A new species of *Dicopomorpha* (Hymenoptera: Mymaridae) with diminutive, apterous males. *Annals of the Entomological Society of America.* 90:115–120. doi.org/10.1093/aesa/90.2.115

For an overview of insect miniaturization, revel in:

Polilov AA. 2015. Small is beautiful: Features of the smallest insects and limits to miniaturization. *Annual Review of Entomology.* 60:103–121. doi.org/10.1146/annurev-ento-010814-020924

flying through the air . . . is more like swimming:

Some of the smallest insects are still swift, active flyers, as a Vietnamese featherwing beetle (*Paratuposa placentis*) demonstrates: youtube.com/watch?v=X0ziHsPqmJg

Farisenkov SE, et al. 2022. Novel flight style and light wings boost flight performance of tiny beetles. *Nature.* 602:96–100. doi.org/10.1038/s41586-021-04303-7

Her brain volume is almost halved when she transforms into an adult:

Polilov AA. 2012. The smallest insects evolve anucleate neurons. *Arthropod Structure & Development.* 41:29–34. doi.org/10.1016/j.asd.2011.09.001

micro-insects invest up to 16 percent of their body mass to brains:

Eberhard WG, Wcislo WT. 2011. Grade changes in brain-body allometry: Morphological and behavioural correlates of brain size in miniature spiders, insects and other invertebrates. *Advances in Insect Physiology.* 40:155–214. doi.org/10.1016/B978-0-12-387668-3.00004-0

"small animals pay relatively more for absolutely less":

Kilmer JT, Rodríguez RL. 2019. Miniature spiders (with miniature brains) forget sooner. *Animal Behaviour.* 153:25–32. doi.org/10.1016/j.anbehav.2019.04.012

EXOSKELETON ENVY

kabuto escalated in their extravagance:

Su N. 2022. Insects in Japanese culture, and one that saved the country. *American Entomologist.* 68:52–58. doi.org/10.1093/ae/tmac010

large horns and small eyes, or small horns and large eyes:

Emlen DJ. 2001. Costs and the diversification of exaggerated animal structures. *Science.* 291:534–536. doi.org/10.1126/science.1056607

Water condenses on bumps:

Parker AR, Lawrence CR. 2001. Water capture by a desert beetle. *Nature.* 414:33–34. doi.org/10.1038/35102108

Park KC, et al. 2016. Condensation on slippery asymmetric bumps. *Nature.* 531:78–82. doi.org/10.1038/nature16956

Adopting this water-harvesting ability:

Aslan D, et al. 2022. A biomimetic approach to water harvesting strategies: An architectural point of view. *International Journal of Built Environment and Sustainability.* 9:47–60. doi.org/10.11113/ijbes.v9.n3.969

The diabolical ironclad beetle:

Wired: How the "Diabolical" Beetle Survives Being Run Over by a Car, wired.com/story/how-the-diabolical-beetle-survives-being-run-over-by-a-car/

photonic crystals and optical computing:

Galusha JW, et al. 2008. Discovery of a diamond-based photonic crystal structure in beetle scales. *Physical Review E.* 77:050904(R). doi.org/10.1103/PhysRevE.77.050904

a gold tortoise beetle changes color:

MIT News: Toward Printable, Sensor-Laden "Skin" for Robots, news.mit.edu/2017/goldbug-beetle-printable-sensor-laden-skin-robots-0323

Bombardier beetles:

Aneshansley DJ, et al. 1969. Biochemistry at 100°C: Explosive secretory discharge of bombardier beetles (*Brachinus*). *Science.* 165:61–63. doi.org/10.1126/science.165.3888.61

applications for gas turbine igniters:

Beheshti N, McIntosh AC. 2006. A biometric study of the explosive discharge of the bombardier beetle. *International Journal of Design & Nature.* 1:1–9. doi.org/10.2495/D&N-V1-N1-61-69

One of da Vinci's most complicated designs imitates the movement of dragonfly wings:

Museum Leonardo: Mechanical Dragonfly, leonardo3.net/en/l3-works/machines/1473-mechanical-dragonfly.html

the invention of the wheel:

Scholtz G. 2008. Scarab beetles at the interface of wheel invention in nature and culture? *Contributions to Zoology.* 77:139–148. doi.org/10.1163/18759866-07703001

few other organisms engage in wheel-rolling motions:

A strange exception is the golden wheel spider (*Carparachne aureoflava*), which catapults itself down Namibian desert dunes, rolling like a wheel to escape parasitic pompilid wasps at forty-four revolutions per second. Other "wheeling" spiders include a close relative (*C. alba*), and a species of jumping spider. Several insect species also roll to escape predation.

Henschel JR. 1990. Spiders wheel to escape. *South African Journal of Science*. 86:151–152.

robot bearing the artificial eye of a fly:

Franceschini N, et al. 1992. From insect vision to robot vision. *Philosophical Transactions: Biological Sciences*. 337:283–294. doi.org/10.1098/rstb.1992.0106

imitating a compound eye to another level:

Jeong K, et al. 2006. Biologically inspired artificial compound eyes. *Science*. 312:557–561. doi.org/10.1126/science.1123053

Song Y, et al. 2013. Digital cameras with designs inspired by the arthropod eye. *Nature*. 497:95–99. doi.org/10.1038/nature12083

it is what flows across her eye that she uses when communicating distance:

Srinivasan MV, et al. 2000. Honeybee navigation: Nature and calibration of the "odometer." *Science*. 287:851–853. doi.org/10.1126/science.287.5454.851

a compound-eye-inspired hemispherical array:

Floreano D, et al. Miniature curved artificial compound eyes. *Applied Biological Sciences*. 110:9267–9272. doi.org/10.1073/pnas.1219068110

Pericet-Camara R, et al. 2015. An artificial elementary eye with optic flow detection and compositional properties. *Journal of the Royal Society Interface*. 12:20150414. doi.org/10.1098/rsif.2015.0414

***Ormia ochracea* flies have ears:**

Robert D, et al. 1992. The evolutionary convergence of hearing in a parasitoid fly and its cricket host. *Science*. 258:1135–1137. doi.org/10.1126/science.1439820

Robert D, et al. 1998. Tympanal mechanics in the parasitoid fly *Ormia ochracea*: Inter-tympanal coupling during mechanical vibration. *Journal of Comparative Physiology A*. 183:443–452. doi.org/10.1007/s003590050270

with the accuracy matching that of a human:

Mason AC, et al. 2001. Hyperacute directional hearing in a microscale auditory system. *Nature*. 410:686–690. doi.org/10.1038/35070564

fly ears are the source of inspiration:

Miles RN, et al. 2009. A low-noise differential microphone inspired by the ears of the parasitoid fly *Ormia ochracea*. *Journal of Acoustical Society of America*. 125:2013–2026. doi.org/10.1121/1.3082118

Kuntzman ML, Hall NA. 2014. Sound source localization inspired by the ears of the *Ormia ochracea*. Applied Physics Letters. 105:033701. doi.org/10.1063/1.4887370

Thomas Eisner, the clever and curious entomologist:

Eisner carefully and beautifully chronicled his life of discovery in *For Love of Insects*, and revealed the astounding chemical defenses of arthropods in *Secret Weapons: Defenses of Insects, Spiders, Scorpions, and Other Many-Legged Creatures* (with Maria Eisner and Melody Siegler).

Eisner T, Anashansley DJ. 2000. Defense by foot adhesion in a beetle (*Hemisphaerota cyanea*). *Proceedings of the National Academy of Sciences.* 97:6568–6573. doi.org/10.1073/pnas.97.12.6568

Eisner T. 2003. *For Love of Insects.* Harvard University Press. Cambridge, MA. pp. 119–126.

Engineers . . . adopted two lessons from these bristly, oily beetle feet:

Vogel MJ, Steen PH. 2010. Capillarity-based switchable adhesion. *Proceedings of the National Academy of Sciences.* 107:3377–3381. doi.org/10.1073/pnas.0914720107

a toy bulldozer underwater:

Hosoda N, Gorb SN. 2012. Underwater locomotion in a terrestrial beetle: Combination of surface de-wetting and capillary forces. *Proceedings of the Royal Society B.* 279:4236–4242. doi.org/10.1098/rspb.2012.1297

Only two orders of insects have never had wings in their history:

. . . unless you include the three orders of arthropods that, like insects, also have six legs. Their mouthparts are hidden in their head capsules, and most entomologists refer to these orders—Protura, Diplura, and Collembola (springtails)—as "non-insect hexapods."

layers of wing scales that, together, reflect different blues:

Giraldo MA, et al. 2016. Coloration mechanisms and phylogeny of *Morpho* butterflies. *Journal of Experimental Biology.* 219:3936–3944. doi.org/10.1242/jeb.148726

techniques adopting a "hierarchy of disorder":

Song B, et al. 2017. Reproducing the hierarchy of disorder for *Morpho*-inspired, broad-angle color reflection. *Scientific Reports.* 7:46023. doi.org/10.1038/srep46023

The structures causing insect iridescence:

Qualcomm has mimicked Morpho butterfly wings to produce Mirasol's display screens that reflect light rather than transmit it.

LiveScience: Color E-Readers Inspired by Butterflies. livescience.com/5895-color-readers-inspired-butterflies.html

engineers are emulating insects:

Kryuchkov M, et al. 2020. Reverse and forward engineering of *Drosophila* corneal nano-coatings. *Nature.* 585:383–389. doi.org/10.1038/s41586-020-2707-9

Science Briefs: Moth Eye Structure Inspires Glare-resistant Screen Coating. sciencebriefss.com/innovation/moth-eye-structure-inspires-glare-resistant-screen-coating/

folding of ladybird beetle's wings:

Saito K, et al. 2017. Investigation of hindwing folding in ladybird beetles by artificial elytron transplantation and microcomputed tomography. *Proceedings of the National Academy of Sciences.* 144:5624–5628. doi.org/10.1073/pnas.1620612114

Bug Wars:

The New Yorker: The Origami Lab, newyorker.com/magazine/2007/02/19/the-origami-lab

The real-life world champion of folding is the earwig:

"This is something I have been trying to film for years," shares Adrian Smith in his video capturing earwig flight: youtube.com/watch?v=_PNtn6ly9wU

The elegant art of origami finds applications even in our bodies:

ETH Zurich: Earwigs and the Art of Origami, youtube.com/watch?v=oNQ_nn3VLiY

ethz.ch/en/news-and-events/eth-news/news/2018/03/earwigs-and-the-art-of-origami.html

Faber JA, et al. 2018. Bioinspired spring origami. *Science*. 359:1386–1391. doi.org/10.1126/science.aap7753

Eaton-Evans J, et al. 2008. Design of an origami stent graft. *Proceedings of the International Conference on Shape Memory and Superelastic Technologies*. pp. 513–520.

Kato is featured in a book of origami masters:

Shuki Kato shares some of his folding techniques in a book about origami insects, including spraying water on tracing paper, one section at a time, and using the thinnest paper available.

Gerstein S. 2013. *Origami Masters: Bugs: How the Bug Wars Changed the Art of Origami*. Race Point Publishers. New York, NY.

the [clanger cicada] wing can destroy many incoming bacteria:

Pogodin S, et al. 2013. Biophysical model of bacterial cell interactions with nano-patterned cicada wing surfaces. *Biophysical Journal*. 104:835–840. doi.org/10.1016/j.bpj.2012.12.046

3D-printed a wrist splint:

Khaheshi A, et al. 2021. Spiky-joint: A bioinspired solution to combine mobility and support. *Applied Physics A*. 127:18. doi.org/10.1007/s00339-021-04310-5

Vulcain released the Cricket Calibre:

MasterHorologer: Vulcain Cricket Calibres—A Look Back. masterhorologer.com/2015/09/12/vulcain-cricket-calibres-a-look-back/

invention of the flexible catheter originating with . . . Benjamin Franklin:

Hirschmann JV. 2005. Benjamin Franklin and medicine. *Annals of Internal Medicine*. 143:830–834. doi.org/10.7326/0003-4819-143-11-200512060-00012

the penis of the thistle tortoise beetle:

Matsumura Y, Kovalev AE, Gorb SN. 2017. Penetration mechanics of a beetle intro-mittent organ with bending stiffness gradient and a soft tip. *Science Advances*. 3:eaao5469. doi.org/10.1126/sciadv.aao5469

steerable microcatheter . . . the thistle tortoise beetle penis:

Gopesh T, et al. 2021. Soft robotic steerable microcatheter for the endovascular treatment of cerebral disorders. *Science Robotics*. 6:eabf0601. doi.org/10.1126%2Fscirobotics.abf0601

Ovipositors have the ability to smell:

Yadav P, Borges RM. 2017. The insect ovipositor as a volatile sensor within a closed microcosm. *Journal of Experimental Biology*. 220:1554–1557. doi.org/10.1242/jeb.152777

steerable needle after ovipositors of parasitoid wasps:

Scali M, et al. 2018. Ovipositor-inspired steerable needle: Design and preliminary experimental evaluation. *Bioinspiration & Biomimetics.* 13:016006. doi.org/10.1088/1748-3190/aa92b9

ovipositor model . . . neurosurgery:

Secoli R, et al. 2022. Modular robotic platform for precision neurosurgery with a bioinspired needle: System overview and first in-vivo deployment. *PLoS ONE.* 17:eabf0601. doi.org/10.1371/journal.pone.0275686

"Taking as his model the sting of the bee, he had constructed a small [glass] syringe":

Brown, T. 1886. *Alexander Wood, M.D., F.R.C.P.E. &c. &c.: A Sketch of his Life and Work.* Macniven & Wallace. Edinburgh, UK. pp. 108–109.

Ellis H. 2017. Alexander Wood: Inventor of the hypodermic syringe and needle. *British Journal of Hospital Medicine.* 78:647. doi.org/10.12968/hmed.2017.78.11.647

spiny genitalia as "pseudo-stings":

Sugiura S, Tsujii M. 2022. Male wasp genitalia as an anti-predator defense. *Current Biology.* 32:R1336–R1337. doi.org/10.1016/j.cub.2022.11.030

a longhorn beetle . . . has a venom-injecting stinger:

Berkov A, et al. 2008. Convergent evolution in the antennae of a cerambycid beetle, *Onychocerus albitarsis,* and the sting of a scorpion. *Naturwissenschaften.* 95:257–261. doi.org/10.1007/s00114-007-0316-1

hypodermic needles that mimic . . . the mosquito's proboscis:

Aoyagi S, et al. 2012. Equivalent negative stiffness mechanism using three bundled needles inspired by mosquito for achieving easy insertion. *IEEE/RSJ International Conference on Intelligent Robots and Systems.* doi.org/10.1109/IROS.2012.6386088

Li ADR, et al. 2020. Mosquito proboscis-inspired needle insertion to reduce tissue deformation and organ displacement. *Scientific Reports.* 10:12248. doi.org/10.1038/s41598-020-68596-w

RISE OF THE INSECT ROBOTS

transforming an agile escape artist into a compliant feast for her young:

Piek T, et al. 1984. Change in behaviour of the cockroach, *Periplaneta americana,* after being stung by the sphecid wasp *Ampulex compressa. Entomologia Experimentalis et Applicata.* 35:195–203. doi.org/10.1111/j.1570-7458.1984.tb03379.x

Herzner G, et al. 2013. Larvae of the parasitoid wasp *Ampulex compressa* sanitize their host, the American cockroach, with a blend of antimicrobials. *Biological Sciences.* 110:1369–1374. doi.org/10.1073/pnas.1213384110

Arvidson R, et al. 2018. Life history of the emerald jewel wasp *Ampulex compressa. Journal of Hymenoptera Research.* 63:1–13. doi.org/10.3897/jhr.63.21762

design robots that mimic cockroaches:

Li C, et al. 2015. Terradynamically streamlined shapes in animals and robots enhance traversability through densely cluttered terrain. *Bioinspiration & Biomimetics.* 10:046003. doi.org/10.1088/1748-3190/10/4/046003

artificial water striders:

Hu D, et al. 2003. The hydrodynamics of water strider locomotion. *Nature*. 424:663–666. doi.org/10.1038/nature01793

Hu D. 2020. *How to Walk on Water and Climb up Walls: Animal Movement and the Robots of the Future*. Princeton University Press. Princeton, NJ.

Koh J, et al. 2015. Jumping on water: Surface tension-dominated jumping of water striders and robotic insects. *Science*. 349:517–521. doi.org/10.1126/science.aab1637

Science: Video: Tiny Robot Walks, Jumps on Water. sciencemag.org/news/2015/07/video-tiny-robot-walks-jumps-water

inchworm robot:

Yang M, et al. 2021. Inchworm-inspired soft robotic climber with embedded fiber-optic sensors. *Optical Fiber Communications Conference*. San Francisco, CA. doi.org/10.1364/OFC.2021.Tu6C.3

Zeng X, et al. 2021. Theoretical and experimental investigations into a crawling robot propelled by piezoelectric material. *Micromachines*. 12:1577. doi.org/10.3390/mi12121577

Khan MB, et al. 2020. iCrawl: An inchworm-inspired crawling robot. *IEEE Access*. 8:200655–200668. doi.org/10.1109/ACCESS.2020.3035871

leaderless, but collective, problem-solving robots:

Krieger MJ, et al. 2000. Ant-like task allocation and recruitment in cooperative robots. *Nature*. 406:992–995. doi.org/10.1038/35023164

Garnier S, et al. 2013. Do ants need to estimate the geometrical properties of trail bifurcations to find an efficient route? A swarm robotics test bed. *PLoS Computational Biology*. 9:e1002903. doi.org/10.1371/journal.pcbi.1002903

Liu T, et al. 2021. A multiple pheromone communication system for swarm intelligence. *IEEE Access*. 9:148721–148737. doi.org/10.1109/ACCESS.2021.3124386

The dragonfly anticipates a prey item's flight trajectory:

Mischiati M, et al. 2015. Internal models direct dragonfly interception steering. *Nature*. 517:333–338. doi.org/10.1038/nature14045

Olberg RM, et al. 2000. Prey pursuit and interception in dragonflies. *Journal of Comparative Physiology A*. 186:155–162. doi.org/10.1007/s003590050015

a computer chip that . . . accomplishes what a brain does:

IEEE Spectrum: Fast, Efficient Neural Networks Copy Dragonfly Brains. spectrum.ieee.org/fast-efficient-neural-networks-copy-dragonfly-brains

Francis Crick . . . would be pleased with this approach:

Crick F. 1998. *What Mad Pursuit*. Sloan Foundation Science Series. New York, NY. pp. 114–115.

MantisBot:

Szczecinski NS, et al. 2015. Introducing MantisBot: Hexapod robot controlled by a high-fidelity, real-time neural simulation. *IEEE IROS*. pp. 3875–3881. doi.org/10.1109/IROS.2015.7353922

This fly's connectome:
Winding M, et al. 2023. The connectome of an insect brain. *Science.* 379:eadd9330.
    doi.org/10.1126/science.add9330
Insectothopter:
Central Intelligence Agency: Artifacts—Insectothopter. cia.gov/legacy/museum/artifact/
    insectothopter/
The Black Vault. documents2.theblackvault.com/documents/cia/insectothopter-cia1.pdf
The race is on to produce the first of a fleet of robots:
Savage N. 2015. Aerodynamics: Vortices and robobees. *Nature.* 521:S64–S65.
    doi.org/10.1038/521S64a
insect robots have grown (as they've shrunk) more sophisticated:
Chen C, Zhang T. 2019. A review of design and fabrication of the bionic flapping wing
    micro air vehicles. *Micromachines.* 10:144. doi.org/10.3390/mi10020144
Tanaka H, Shimoyama I. 2010. Forward flight of swallowtail butterfly with simple
    flapping motion. Bioinspiration and Biomimetics. 5:(2010)026003.
    doi.org/10.1088/1748-3182/5/2/026003
YouTube: Insect-Sized Robot Takes Flight: RoboBee X-Wing. youtube.com/
    watch?v=loHzoeFP9Io
YouTube: Insect-Like Robots. youtube.com/watch?v=50_kK9phHy8
designs to hijack actual insects:
Pulla S, Lal A. 2009. Insect powered micro air vehicles. *2009 IEEE International
    Conference on Robotics and Automation.* pp. 3657–3662. doi.org/10.1109/
    ROBOT.2009.5152878
others have inserted electrodes:
Sato H, Maharbiz MM. 2010. Recent developments in the remote radio control of insect
    flight. *Frontiers in Neuroscience.* 4. doi.org/10.3389/fnins.2010.00199
your very own RoboRoach Bundle:
Backyard Brains: The RoboRoach Bundle, backyardbrains.com/products/roboroach
Should they wish for a fly to take off or land, they simply flash a light:
Centre for Neural Circuits and Behaviour: Gero Miesenböck FRS. cncb.ox.ac.uk/people/
    gero-miesenboeck/
cyborg dragonflies:
Popular Mechanics: This Genetically-Modified Cyborg Dragonfly Is the Tiniest Drone.
    popularmechanics.com/flight/drones/a26729/genetically-modified-cyborg-
    dragonfly/
Draper: Equipping Insects for Special Service. draper.com/news-releases/equipping-
    insects-special-service
Janelia Research Campus: Leonardo Lab, janelia.org/lab/leonardo-lab

ART, AND FABRICATING THE PERFECT INSECT

plastic cockroaches make for wholesome, heart-stopping fun:

One review of Wirrabilla's Premium Fake Cockroaches purchased online: "Boy did I enjoy these! After placing them on the bathroom floor I faked picking them all up then threw them on my girlfriend. I've never laughed so hard! Anyway she's available now . . ." (To date, two people have found this review to be helpful.)

A second review, from Hapeville, Georgia: "Really good in Yoplait Original Yogurt, Strawberry and Strawberry Banana, Variety Pack, 12 Cups!"

Haruo Mitsuta had thought the practice extinct for a century:

Haruo Mitsuta & Zizai Specimen Box, m-haruo.com/index.html

YouTube: How to Make the Metal Stag Beetle, youtube.com/watch?v=Jfa9jJM3npA

Interview, www3.nhk.or.jp/nhkworld/en/ondemand/video/2058951

Leather beetle pouches and flea bags:

amaheso.com

fruit leather beetles:

The New York Times Magazine: Noma, Rated the World's Best Restaurant, Is Closing Its Doors, nytimes.com/2023/01/09/dining/noma-closing-rene-redzepi.html

the model of a male mosquito:

Lucas FA. 1936. *General Guide to the Exhibition Halls of the American Museum of Natural History*, 21st ed. American Museum of Natural History. New York, NY. p. 29.

Dahlgren BE. 1908. *The Malaria Mosquito: A Guide Leaflet Explanatory of a Series of Models in the American Museum of Natural History* (Guide leaflet no. 27). American Museum of Natural History, New York, NY. digitallibrary.amnh.org/handle/2246/7166

insect masterpieces:

Google Arts & Culture: Keller's Insect Tales. artsandculture.google.com/story/GgWxfqxgzrPYIw

A fly is covered in 2,653 bristles:

Kemp M. 2010. Sculpture: Terrible wonder. *Nature*. 468:506–507. doi.org/10.1038/468506a

Auzoux's papier-mâché:

Whipple Museum of the History of Science: Animal Models. whipplemuseum.cam.ac.uk/explore-whipple-collections/models/dr-auzouxs-papier-mache-models/animal-models

Physical models expand a scientist's toolbox:

Klein BA, et al. 2012. Robots in the service of animal behavior. *Communicative and Integrative Biology*. 5:466–472. doi.org/10.4161/cib.21304

silicone molds of real caterpillars:

Papaj DR, Newsom GM. 2005. A within-species warning function for an aposematic signal. *Proceedings of the Royal Society B*. 272:2519–2523. doi.org/10.1098/rspb.2005.3186

bumblebees learn . . . by watching other bumblebees:
Worden BD, Papaj DR. 2005. Flower choice copying in bumblebees. *Biology Letters.* 22:504–507. doi.org/10.1098/rsbl.2005.0368

[Bats] preferred aluminum dragonflies with crumply wings:
Geipel I, et al. 2013. Perception of silent and motionless prey on vegetation by echolocation in the gleaning bat *Micronycteris microtis. Proceedings of the Royal Society B.* 280:20122830. doi.org/10.1098/rspb.2012.2830

males chose the paper models with smaller spots:
Fordyce JA, et al. 2002. The significance of wing pattern diversity in the Lycaenidae: Mate discrimination by two recently diverged species. *Journal of Evolutionary Biology.* 15:871–879. doi.org/10.1046/j.1420-9101.2002.00432.x

Beetles flew and landed on the bioreplicated decoys:
Domingue MJ, et al. 2014. Bioreplicated visual features of nanofabricated buprestid beetle decoys evoke stereotypical male mating flights. *Applied Biological Sciences.* 111:14106–14111. doi.org/10.1073/pnas.1412810111

ARCHITECTURE: BUILDING WHAT THEY BUILD

Aluminum cast of a nest of fire ants:
Walter Tschinkel has mastered the art of casting and excavating elaborate ant nests to learn about the ecology and architecture of ants.
Tschinkel WR. 2015. The architecture of subterranean ant nests: Beauty and mystery underfoot. *Journal of Bioeconomics.* 17:271–291. doi.org/10.1007/s10818-015-9203-6
Tschinkel WR. 2021. *Ant Architecture.* Tantor and Blackstone Publishing. Old Saybrook, CT.

Evolutionary Architecture:
Tssui, E. 1999. *Evolutionary Architecture: Nature as a Basis for Design* 1st ed. Wiley. Hoboken, NJ.

termite mounds act much more like a human lung:
Turner S, Soar RC. 2008. Beyond biomimicry: What termites can tell us about realizing the living building. *Proceedings of 1st International Conference on Industrialized, Intelligent Construction.*

Sustainable architecture is achievable, and insects have the answers:
Gorb SV, Gorb EV. 2020. Insect-inspired architecture to build sustainable cities. *Current Opinion in Insect Science.* 40:62–70. doi.org/10.1016/j.cois.2020.05.013

Many buildings . . . are modeled after hives humans have designed:
Prendergast KS, et al. 2021. Bee representations in human art and culture through the ages. *Art & Perception.* 10:1–62. doi.org/10.1163/22134913-bja10031

building robots that act somewhat like . . . termites, collectively constructing buildings:
Werfel J, et al. 2014. Designing collective behavior in a termite-inspired robot construction team. *Science.* 343:754–758. doi.org/10.1126/science.1245842

a building in the desert that passively collects water:
Aslan D, et al. 2022. A biomimetic approach to water harvesting strategies: An architectural point of view. *International Journal of Built Environment and Sustainability.* 9:47–60. doi.org/10.11113/ijbes.v9.n3.969

FIGHT LIKE AN INSECT

BOOKANIMA: *Praying Mantis*:

Ting L. 1999. *Seven-Star Praying Mantis Kung Fu*. International Wing Tsun Leung Ting Martial Art Association.

Shon Kim's other book animations: shonkim.com/bookanima-2018

***A Lot of Insects*:**

Frank Lutz collected from the "lot" of his yard to demonstrate that "more different kinds of insects either live in or come of their own free will to the 75 × 200-foot yard of my home in the middle of a suburban town than there are different kinds of birds in all of the United States and Canada."

Lutz FE. 1941. *A Lot of Insects: Entomology in a Suburban Garden*. G. P. Putnam's Sons. New York, NY. p. 86.

Wang Lang was planning to incite a rebellion:

Chao HC. 1986. *Chinese Praying Mantis Boxing*. 2nd ed. Unitrade Co. Ltd. China.

Wang Lang took the mantis home; combatting opponents' advances:

Ming TW, Sang Y. 1991. *Taichi Mantis Volley-catch Boxing*, 2nd ed. Yih Mei Book Co., Ltd. Wanchai.

animating the legs with hops, low kicks, and sweeps:

Kahn L. 1988. Crushing kicks of praying mantis: Kung fu's cold-blooded killer. *Black Belt*. 26:20–25.

DANCE LIKE AN INSECT

The Ju|'hoansi women form a line that bobs and undulates:

van Huis A. 2019. Cultural significance of Lepidoptera in sub-Saharan Africa. *Journal of Ethnobiology and Ethnomedicine*. 15:26. doi.org/10.1186/s13002-019-0306-3

YouTube: Zu/'hoasi Women in the Caterpillar Dance, youtube.com/watch?v=6Z8D-N9Dhg4

Megan Biesele wrote the following to me: "I have spent a total of ten years over the course of five decades with the Ju|'hoansi (new orthography they have adopted) and have never had the privilege of seeing a Caterpillar Dance. I still learn something wonderful and new every single day about these people. My guess is that the Caterpillar is one of many, more casual, animal-imitation dances of the Ju|'hoansi."

The Bug:

Library of Dance: The Bug, libraryofdance.org/dances/bug

"the frenzy of jittering bugs":

DanceFans: Jitterbug and Lindy Hop, dancefans.cultu.be/jitterbug-and-lindy-hop

Polik-mana, the Butterfly Maiden spirit:

YouTube: Hopi Festival 2018, youtube.com/watch?v=jFG1Uj-c-yM

Circle of Dance: Hopi Butterfly Dance. americanindian.si.edu/static/exhibitions/circleofdance/hopi.html

a mountain on Venus bears her name:

Gazetteer of Planetary Nomenclature: Polik-mana Mons. planetarynames.wr.usgs.gov/Feature/4780

fancy shawl dance:

ICT: The Evolving Beauty of the Fancy Shawl Dance. ictnews.org/archive/the-evolving-beauty-of-the-fancy-shawl-dance

Women imitate butterflies . . . children mimicking the moth that brings sleep; The mask whirls like a propeller to activate a rain charm:

Lonsdale S. 1981. *Animals and the Origins of Dance*. Thames & Hudson. New York, NY. pp. 141–142; 22, 106.

praying mantis dance:

Into Cambodia: Robam Kandob Ses (Praying Mantis Dance). intocambodia.org/content/robam-kandob-ses-praying-mantis-dance

YouTube: Robam Kandob Ses (Praying Mantis Dance). youtube.com/watch?v=SSCSsYLQ6u0

kocho (butterfly) dance:

YouTube: Takekiri-eshiki, youtube.com/watch?v=MPnHB23gF38

the Ainu of Hokkaido imitate grasshoppers:

YouTube: Traditional Ainu Dance (last dance; 8 minutes 2 seconds). youtube.com/watch?v=bKaw7ShYIYQ

A Filipino folk dance called the alitaptap:

YouTube: ALITAPTAP Fold Dance, youtube.com/watch?v=uKlTnyB4GRc

Australia [has] a wealth of danceable insect biodiversity:

Lonsdale S. 1981. *Animals and the Origins of Dance*. Thames & Hudson. New York, NY. Green hornet boring hole: p. 77; caterpillar dancers: pp. 114–115.

honey ant . . . ceremony:

Spencer B, Gillen FJ. 1899. *The Native Tribes of North Central Australia*, sacred-texts.com/aus/ntca/ntca08.htm

spencerandgillen.net/

Mushi no Hoshi—Space Insect:

The Georgia Straight: Mushi no Hoshi Resplices our Dark Future. straight.com/arts/407726/mushi-no-hoshi-resplices-our-dark-future

Vimeo: Dairakudakan—Mushi no Hoshi—Space Insect (excerpts), vimeo.com/115778433

Merce Cunningham studied insects under microscopes:

The New York Times Magazine: Merce Cunningham the Maverick of Modern Dance. nytimes.com/1982/03/21/magazine/merce-cunningham-the-maverick-of-modern-dance.html

Paul Taylor . . . arranging collages of insects:

The Washington Post: Paul Taylor Prolific Modern Dance Choreographer Dies at 88. washingtonpost.com/local/obituaries/paul-taylor-prolific-modern-dance-choreographer-dies-at-88/2018/08/30/fd8c5fc4-ac60-11e8-8a0c-70b618c98d3c_story.html

"She didn't ever play human":

Jowitt D. 2004. *Jerome Robbins: His Life, His Theater, His Dance*. Simon & Schuster. New York, NY. p. 190.

Fuller forged a friendship with Marie Curie:
Ars Technica: Two visionaries: Marie Curie forged a friendship with dancer Loïe Fuller. arstechnica.com/science/2021/02/when-physics-met-dance-marie-curie-and-loie-fuller-in-belle-epoque-paris/
inspired Jody Sperling's and Elizabeth Aldrich's *The Butterfly Dance*:
YouTube: The Butterfly Dance, Performed by Jody Sperling. youtube.com/watch?v=VrzdeSkYp5Q
Julia Oldham hops and flutters while barefoot in the woods:
University of Chicago Magazine: A Bug's Dance. magazine.uchicago.edu/0912/arts_sciences/dance.shtml
Vimeo: The Timber, 2009, vimeo.com/171691978
created night by piercing the gallbladder of a slain antelope:
Another arthropod origin of night comes from the Andaman Islands, India. Da Tengat, a spider created by the spider goddess Biliku, crushed a cicada, and the shrill cry of the cicada pierced the air and darkened the sky for the first time. It took torches, song, and dance to restore light in the form of dawn.
Lonsdale S. 1981. *Animals and the Origins of Dance*. Thames & Hudson. New York, NY. pp. 50–51.
"The 'butterfly' is danced with bent knees . . .":
Dance Hall: Embodied Dance Hall Choreographies: 'Dance wi a dance an' a bun out a . . .'? dancehallgeographies.wordpress.com/tag/butterfly-dance/
YouTube: How To Do the Butterfly, youtube.com/watch?v=rKf6jsYlnN8
Sidorova first imitated a spider in her living room at thirteen years old:
Milena Sidorova: *The Spider*, milenasidorova.com/spider
Sidorova modeled for a humanoid spider in the film *Dune*: Screen Rant: *Dune's* Spider Monster Explained By VFX Artists, screenrant.com/dune-2021-spider-monster-vfx-explained-video/

ACT LIKE AN INSECT
"it would be a major Aesopian error to believe otherwise":
Charlotte Sleigh shares this quote in: Sleigh C. 2007. *Six Legs Better: A Cultural History of Myrmecology*. Johns Hopkins University Press. Baltimore, MD. p. 225.
Original source: "The Coherence of Knowledge," Phi Beta Kappa oration given at Harvard University, 2 June 1998, Wilson papers.
attempted to apply lessons learned from the bees to his human colleagues:
Seeley TD. 2011. *Honeybee Democracy*. Princeton University Press. Princeton, NJ.
economies often behave like living organisms:
Ormerod P. 1998. *Butterfly Economics: A New General Theory of Social and Economic Behavior*. Faber and Faber Limited. London, UK.
Just observe leaf-cutter ants:
Dussutour A, et al. 2009. Priority rules govern the organization of traffic on foraging trails under crowding conditions in the leaf-cutting ant *Atta colombica*. *Journal of Experimental Biology*. 212:499–505. doi.org/10.1242/jeb.022988

Farji-Brener AG, et al. 2010. The "truck-driver" effect in leaf-cutting ants: How individual load influences the walking speed of nestmates. *Physiological Entomology*. 36:128–134. doi.org/10.1111/j.1365-3032.2010.00771.x

Ants can solve difficult optimization problems:

The Argentine ant can solve a dynamically changing Towers of Hanoi puzzle, as shown here: Reid CR, et al. 2011. Optimisation in a natural system: Argentine ants solve the Towers of Hanoi. *Journal of Experimental Biology*. 214:50–58. doi.org/10.1242/jeb.048173

evidence supporting the biological basis for cooperation:

Allee WC. 1943. Where angels fear to tread: A contribution from general sociology to human ethics. *Science*. 97:517–525. doi.org/10.1126/science.97.2528.517

"If you want peace":

Originally published here:

Forel A. 1928. *The Social World of the Ants: Compared with that of Man*. G.P. Putnam's, London. pp. xix–xx. archive.org/details/dli.ministry.06393/page/n1/mode/2up

and placed in a larger context here: Rodgers DM. *Debugging the Link between Social Theory and Social Insects*. LSU Press. Baton Rouge, LA. p. 166.

An insect society can represent ideals attributed to a monarchy:

James Costa writes about what we see, as humans, in a social insect colony: Costa J. 2002. Scale models? What insect societies teach us about ourselves. *Proceedings of the American Philosophical Society*. 146:170–180.

debate as to how to translate "*ungeheueres Ungeziefer*":

Roughly "monstrous vermin," Franz Kafka informed his publisher that he did not want the "Insekt" to be pictured on the cover. Some have turned Gregor Samsa into a beetle, and the story's cleaning lady specifically calls Samsa a dung beetle ("Mistkäfer").

The New Yorker: On Translating Kafka's *The Metamorphosis*. newyorker.com/books/page-turner/on-translating-kafkas-the-metamorphosis

"insect exercise":

Cassiers E, et al. 2015. Physiological performing exercises by Jan Fabre: An additional training method for contemporary performers. *Theatre, Dance and Performance Training*. 6:273–290. doi.org/10.1080/19443927.2015.1084846

Lindemann A. 2022. The Colony: An evo devo art performance on social life. *Leonardo*. 55:338–344. doi.org/10.1162/leon_a_02222

***The Colony*:**

Anna Lindemann, annalindemann.com/thecolony

DRESS LIKE AN INSECT

Moth eyebrows:

History Collection: Looks that Kill: Impossible Beauty Standards from History. historycollection.com/changing-face-fashion-11-historical-standards-beauty/3

Newhanfu: History of Traditional Chinese Eyebrows Makeup, newhanfu.com/24960.html

SOUND LIKE AN INSECT

A beautifully preserved pair of fossilized wings:

Gu J, et al. 2012. Wing stridulation in a Jurassic katydid (Insecta, Orthoptera) produced low-pitched musical calls to attract females. *Applied Biological Sciences*. 109:3868–3873. doi.org/10.1073/pnas.1118372109

in our music, on album covers, and in music videos:

Rothenberg D. 2013. *Bug Music: How Insects Gave us Rhythm and Noise*. Picadore. London, UK.

For example, rock:

Coelho JR. 2000. Insects in rock and roll music. *American Entomologist*. 46:186–200. doi.org/10.1093/ae/46.3.186

Coelho JR. 2004. Insects in rock and roll cover art. *American Entomologist*. 50:142–151. doi.org/10.1093/ae/50.3.142

Coelho JR. 2021. Sex, bugs and rock 'n' roll: Insects in music videos. *Insects*. 12:616. doi.org/10.3390/insects12070616

cicada imitations:

Versions of the Dong (Kam) and T'boli imitations of cicadas can be found here: Folklife: Cicada Folklore, or Why We Don't Mind Billions of Burrowing Bugs at Once. folklife.si.edu/magazine/cicada-folklore

***Voyagers 1* and *2* each carry a golden record; "vociferous life on earth":**

Sagan C, et al. 1978. *Murmurs of Earth*. Random House, New York, NY.

"There will be a last sentient being":

Quoted by Dennis Overbye in this moving opinion piece: *The New York Times*: Who Will Have the Last Word on the Universe? nytimes.com/2023/05/02/science/end-of-universe.html

CONCLUSION

"May this tusk root out the lice":

Vainstub D, et al. 2022. A Canaanite's wish to eradicate lice on an inscribed ivory comb from Lachish. *Jerusalem Journal of Archaeology*. 2:76–119. doi.org/10.52486/01.00002.4

Garfinkel, quoted in *The New York Times*: nytimes.com/2022/11/09/science/ivory-comb-beard-lice.html

changed the course of history:

Chaline E. 2011. *Fifty Animals that Changed the Course of History*. Firefly Books. Buffalo, NY.

inspirations for engineers:

Fournier M. 2011. *Biomimétisme: Quand la nature inspire la Science*. Plume de carotte. Toulouse, France.

each of us can change the world:

Kawahara AY, et al. 2021. Eight simple actions that individuals can take to save insects from global declines. *Proceedings of the National Academy of Sciences*. 118:e2002547117. doi.org/10.1073/pnas.2002547117

Forister ML, et al. 2019. Declines in insect abundance and diversity: We know enough to act now. *Conservation Science and Practice*. 1:e80. doi.org/10.1111/csp2.80

Barrett:

When I asked Syd Barrett's nephew, Ian Barrett, about his uncle's choice of insects and intention for the image, Ian responded, "As a family, we are all pretty obsessed with the natural world and science in many ways so it is pretty obvious really. I suspect he just loved the beauty of the insects and wanted to capture them. Whether it was ever meant to be an album cover I doubt."

Notes:

Antennae and leg spurs of the Caribbean cockroach inspired this typeface, insectile, by P22 Type Foundry.

Dear Climate (2014–ongoing), courtesy of Marina Zurkow and Dear Climate

INDEX

Portrait of the author by Karen Anne Klein (2013)

BARRETT KLEIN investigates mysteries of sleep in societies of insects, creates entomo-art, and is ever on the search for curious connections that bind our lives with our six-legged allies. Barrett studied entomology at Cornell University and the University of Arizona, fabricated natural history exhibits at the American Museum of Natural History, worked with honey bees for his PhD at the University of Texas at Austin, and spearheaded the Pupating Lab at the University of Wisconsin–La Crosse. He celebrates biodiversity and the intersection of science and art, and believes fully that embracing the beauty of insects can transform our lives and our world.